Visualization and Interpretation

Visualization and Interpretation

Humanistic Approaches to Display

Johanna Drucker

The MIT Press
Cambridge, Massachusetts
London, England

This book was set in Stone Serif and Stone Sans by Westchester Publishing Services. Printed and bound in the United States of America.

Library of Congress Cataloging-in-Publication Data

Names: Drucker, Johanna, 1952- author.
Title: Visualization and interpretation : humanistic approaches to
    display / Johanna Drucker.
Description: Cambridge, Massachusetts : The MIT Press, 2020. |
    Includes bibliographical references and index.
Identifiers: LCCN 2020002673 | ISBN 9780262044738 (hardcover)
Subjects: LCSH: Visual communication. | Graphic arts. | Visual analytics.
Classification: LCC P93.5 .D85 2020 | DDC 302.2/26--dc23
LC record available at https://lccn.loc.gov/2020002673

10 9 8 7 6 5 4 3 2 1

Contents

Acknowledgments

Some sections of this work draw on materials from the author's previously published work. All are republished or cited either with explicit permission or in accord with the authors' rights guidelines:

"Humanities Approaches to Interface Theory," *Culture Machine* 12 (2011) (by permission).

"Humanistic Theory and Digital Scholarship," in *Debates in Digital Humanities*, ed. Matthew Gold (Minneapolis: University of Minnesota Press, 2012), pp. 85–95.

"Non-representational Approaches to Modelling Interpretation in a Graphical Environment," *Digital Scholarship in the Humanities* 33, no. 2 (June 2018) (by permission and guidelines), https://doi.org/10.1093/llc/fqx034 (5300).

"Information Visualization and/as Enunciation," *Journal of Documentation* 73, no. 5 (2017), pp. 903–916, https://doi.org/10.1108/JD-01-2017-0004.

"Performative Materiality and Theoretical Approaches to Interface," *DHQ* (*Digital Humanities Quarterly*) 7, no. 1 (Summer 2013).

"Design Agency," *Dialectic* 1, no. 2 (2017), pp. 11–16.

"Digital Ontologies: The Ideality of Form in/and Code Storage—or—Can Graphesis Challenge Mathesis?," *Leonardo* 34, no. 2 (2001), pp. 141–145.

All images are property of Johanna Drucker.

Framework: Creating the Right Tools and Platforms

In the several decades since humanists have taken up computational tools, they have borrowed many techniques from other fields. This has included the appropriation of visualization methods to create charts, graphs, diagrams, maps, and other graphic displays of information. But are these visualizations actually adequate for the interpretative approach that distinguishes much of the work in the humanities? To answer that question, we have to define the features of that interpretative work and identify the ways its assumptions and premises are distinct from those of other research methods. We also have to ask what kinds of attitudes toward knowledge can be expressed with current conventions of visualization and consider alternatives. If, as will be argued here, the activity of *modeling interpretation* is fundamentally at odds with current methods of *information visualization* (data display), then how would such interpretative work be structured within a computational environment and a user interface? The challenges are many.

We can start by considering what is being referred to in the phrase *information visualization* and how it contrasts with that of *modeling interpretation*. The standard approach to information visualization is to generate a graphic from live or static data. The process is relatively straightforward. A set of quantitative values is charted on a grid, plane, or space governed by a regular, standard metric. Columns, bars, pie charts, points, lines, network diagrams, and so on are all produced in this way. The interpretative work of shaping the data disappears from view in the final result. The image displayed on screen, in print, or through other output devices appears as a statement of fact. The interpretative dimensions of the activity that shaped the data are rendered invisible, not so much concealed as simply missing from view, absent without a trace.

Any statistician or quantitative analyst will freely admit to the interpretative aspects of data creation (choices about sample size, parameterization, and other aspects of statistical and quantitative manipulations).[1] But the idea that the graphic display is a *presentation of the data* stands unquestioned. No additional consideration is necessary. The image is considered to be the data expressed in graphic form. The goal is to find the "best" and "clearest" display of data, and the correlation between the data set and the presentation is assumed to be direct, a matter of equivalence.[2] Visualizations are not characterized, for instance, as remediations (translations of documents or objects from analog to digital format). They are not described as the outcome of interpretative parameters (methods of quantification) that translate numerical values into graphical ones. These methods of translation use conventions that may or may not be logically consistent with the display of quantitative information (the scale or measure may have been altered). And often, the process of data creation produces results that are far from the original phenomenon from which it was extracted. But these processes of transformation—from phenomenon to data and then to display—are rarely documented or noted. Instead, the output is simply called a "data visualization," as if the process were unproblematic and as if the data and visualization were one and the same. In general, this is a one-way process, with visualizations generated from (that is, *after* the creation of) a data set. Sometimes values can be added to the underlying data through an interface (even the language used here emphasizes the notion that data exist first, as a primary layer, to which any modification is a later addition or layer). The data *structure* is almost never created or modified through direct manipulation of the graphical features in this situation—only the most superficial features of display (such as color or use of symbols) can be changed.

The idea of *modeling interpretation*, proposed here, offers a clear contrast to these practices of data visualization. The idea of *modeling* implies that a graphical expression serves as a primary mode of knowledge production, not a secondary expression of preexisting data. This suggests that a graphical expression might be used to create and/or show the features of a model of interpretative activity. This modeling approach changes the unidirectional relationship between data and display and turns the process into a two-way exchange. Thus, an act of creating an inflection, adding an attribute, or making a change in a graph would feed back into the structured data. This would create data and could also intervene in the formal structures of the data, even changing its structure. But modeling interpretation also calls for different, distinct graphical forms capable of expressing ambiguity, contradiction, nuance, change, and other aspects of critical consideration. This requires forms and formats that do not conform to the standards of regular metrics on coordinate grids governed by a standard Cartesian system.[3] If we are to model interpretation—instead of displaying information—we have to imagine how this would *look*. We also have to reflect on whether we are creating an environment for *representing* interpretation or *creating* it.

This book addresses these issues from within a humanist perspective informed by interpretative practices (critical hermeneutics) and constructivist approaches to epistemology.[4] Constructivist approaches emphasize the idea that knowledge arises from experience and is not a simple perception of a preexisting world. The constructivist position takes as a premise that objects of attention—whether perceived, read, viewed, experienced—are made through acts of interpretation. (For instance, a fairy tale may exist as a text, but a child hearing this text may experience it as frightening while an adult may find it charming and nostalgic.) In this view, every text (image, artifact, score, etc.) is produced anew in a reading. Every reading proceeds from an attempt at *understanding* through which weave the inevitable and persistent questions of *how* we know what we know.[5] It further suggests that the reader (or reading subject) is changed by every encounter with the text. This reader not only creates a text through the negotiation of a self and a work within a historical horizon of knowledge and condition of understanding, but is also affected by it. We don't read a text literally or mechanically. We interpret it as we read. As a work or text is produced by reading, this experience also affects the reader. Text and reader exist in a relation of reciprocity.

This characterization of reading suggests a codependency that is not merely limited to textual production, but to the production of the reader in the sense suggested in theories of what is termed *enunciation*. In this theoretical framework, which comes from linguistics, the individual reader is produced as a *subject* of a work. Such a subject is spoken, situated, positioned by the work. This effect has implications for the impact of texts within a cultural frame. The approach shifts the ground of a reading experience from that of isolated individual attention to one of collective effect across historical time and a specific space. Just as the architecture of a church *positions* individuals in relation to its structures and rituals, so a text positions readers within its frameworks. Issues of racial bias, gender inequity, power relations, and legacies of colonial practice across categories of identity are structured into the textual systems. The term *enunciation* describes this structuring relationship as it is produced in language. But these concepts also apply to visual systems, as we shall see.

To this familiar formulation of interpretation (or critical hermeneutics), this book adds yet another dimension: the probabilistic nature of reading practices. This approach suggests that readings are interventions within the field of provocation provided by the text. Here, a text (or work of any kind) is considered a provocation, a field of potential or possibility, in which a reading or interpretation is an intervention. The act of intervention creates a unique reading in the provocative field of the work depending on the reader. The range of probability of a reading fits the normal bell curve in many instances—readings of a work will tend to cluster around a consensual understanding.[6] When described this way, the expectation is that the range of readings will follow a normative distribution in line with a fairly mechanistic probability. But a more radical approach to probability and interpretation suggests that the relationship between acts of reading and textual artifacts is always nondeterministic. The act of reading is an intervention in a field of provocation (the text, work, encoded experience). The outcome can be compared to the flattening of a wave function in physics, something that cannot be predicted with certainty. Readings produce a text/work in accord with the specifics of start conditions whose variables cannot be fully described or accounted for in all of their particulars.[7] This probabilistic model of reading will be teased out in more detail in the chapters ahead, along with attention to the issue of *particularity* in its specific relevance to visual forms of knowledge production. The probabilistic model of interpretation described here has implications for visualizations as well as for the data to which they are linked.

Probabilistic models of interpretation are arguments about how knowledge is produced. Thus, when dealing with the themes that will populate this book, such as the variations in systems for constructing chronological time, or for marking historical periods, or for gauging the dimensions of space or narrating encounters of discovery, the problems of modeling the acts of knowing that are central to these themes are considered at least as important as the substance of what is known. In other words, our challenge is not simply to model *what* we know, but *how*—and to recognize that the *what* is always constituted as an effect of the *how*. The task of modeling interpretation begins with these premises. We also have to recognize that the work of display from empiricist, positivist, and mechanistic perspectives is already well served by existing conventions. These have come in large part from the natural sciences and social sciences. In those domains, a minute is always the same, a space can be measured with a standard rule, a narrative unfolds in a linear sequence, and representations are surrogates that stand in stable relation to whatever they represent. By contrast, in a user-dependent hermeneutic approach, the dimensions of time and space are influenced by factors of experience, while any understanding of a text is always produced from a position marked by historical, cultural, and individual factors.

User-dependent knowledge is interpretative, partial, and situated. Therefore, empirical modes of graphic display of information are as unsuited to modeling interpretation as a thermometer is for measuring the warmth of a human emotion or the strength of an embrace. We cannot measure the many (experiential, intellectual, critical) dimensions of knowledge as interpretation with the same tools as those we bring to the study of those phenomena we imagine to be outside of ourselves. In fact, the conclusion I hope to defend through the arguments made here is the contrary: that notions of externalized, user-*in*dependent knowledge are impossible within the critical epistemological approach of the humanities. Further, we might even suggest that this (critical hermeneutic) model has much to offer to fields outside the humanities—and is one of the major contributions the humanities can make to debates about the ideological and affective character of knowledge as *knowing*. The emphasis on *knowing* over *knowledge* substitutes processes (and ongoing negotiations between the subject and object) for a mechanistic model of perception (in which the world is assumed to simply become reified in representations of things).

The emphasis on *knowing* situates the interpretative process in a user-dependent model. The distinction between user-independent and user-dependent models of knowledge distinguishes empirical (mechanistic) from interpretative (hermeneutic) approaches. Within the tenets of modern physics, this line blurs. Since Albert Einstein, Werner Heisenberg, and other physicists working in the early twentieth century began to grapple with the mutable conditions of space-time, the concepts of uncertainty and probability, and the relation between observation and outcome have become part of the study of natural phenomena, particularly at certain scales (such as that of subatomic particles). The humanities have lagged behind. This is changing within the frameworks of probabilistic interpretation (critical hermeneutics), constructivist approaches to knowledge (radical epistemology), and more recently, new materialisms (theories in which the material world has some agency). Humanities scholars have grappled with some of the features of what can be considered *codependence* (between subject and object, reader and text) in the fundamental processes of knowledge production. But these concepts have not been brought to bear on visual display and graphical expressions. Even in textual humanities, a persistent strain of positivism—suggesting that a text has a meaning that can be fixed and grasped—continues to push against the tenets of constructivism, especially in digital projects. The approach to modeling interpretation outlined here builds on established principles of hermeneutic understanding taken from the critical tradition, but makes its contribution through a focus on the specific problems of graphical expression and production of interpretative practices within the probabilistic frameworks just outlined.

Outline of this book

This book is structured in five parts. Chapter 1, "Visual Knowledge (or *Graphesis*)," contains a discussion of visual knowledge production (epistemology) and the specific properties of graphical approaches within a digital environment. Here the questions posed have to do with the relation between inscription and notation, between producing and recording knowledge, and also, with the specificity of visual expressions required for interpretative work. This chapter looks at various theoretical understandings of visual images and their relation to knowledge and asks how the specifics of the graphical are to be engaged directly as a primary means of knowledge production for digital humanities. What kind of knowledge belongs to the visual and how can computational environments support this activity within a critical interpretative (hermeneutic) framework? How do graphical methods produce and inscribe knowledge differently than textual, numeric, or other approaches? The chapter draws on work from aesthetics, critical theory, and formal study of graphical systems and addresses these within the specific framework of computational and digital activity as they apply to digital humanities.

Chapter 2, "Interpretation as Probabilistic," is concerned with defining the specific parameters of the humanistic methods invoked here. It describes what is meant by user-centered conditions of interpretation, how they are enacted, and what purpose is served by outlining a claim for their cultural authority as well as their practical use. Here the tenets of critical interpretation (hermeneutics) meet the actualities of research problems and the discursive heterogeneity of the cultural record. This chapter outlines the basic properties of a partial and situated approach to knowledge as interpretation and its role within the digital humanities. It expands the notion of probabilistic interpretation as a crucial aspect of scholarly work within digital environments. This approach is presented with an argument for its political impact and value, as well as its critical purchase on expressions and instantiations of the cultural record when these are read, analyzed, and studied within the mediated and remediated conditions of digital formats.

Chapter 3, "Graphic Arguments," proposes an approach to graphical methods of interpretation grounded in a nonrepresentational understanding of visualization (beginning with a discussion of what *nonrepresentational* images are). This chapter discusses argument forms and graphical rhetoric as primary tools of and for interpretation. It also proposes a set of conventions for the creation of argument structures and for their direct encoding within data structures. It also addresses the challenge of taking informal approaches, some of which resist the very terms of formal languages and expressions, into structured data. But the more immediate concern of this project is with developing the conventions according to which the rhetorical force of interpretative work can be enacted using graphical methods that link directly to the production of data and code.

Chapter 4, "Interface and Enunciation," takes a step back from the direct work of interpretation (of texts, images, etc.) to address the problems of the enunciative (subject-producing) framework within which digital work takes place and the concealment enacted by interface. The analysis of interface as an enunciative system, and of information as an aspect of that system, is central to understanding the ways in which a reader or scholar's position as a subject is created in screen, device, platform, and projection environments. This chapter also addresses the problems of the lifecycle of data production and its obfuscation. The questions of how data production can be visualized, and how the interpretative modeling of what passes for information can be exposed, are addressed here as part of an attention to the larger problems of the function, performance, and role of the enunciative apparatus of information systems. The link between enunciative systems and the enactment of power relations invokes interface as a major site and instrument of ideology, along with the need to create visual techniques for showing its workings.

Finally, chapter 5, "The Projects in Modeling Interpretation," presents a series of projects that incorporate these theoretical formulations. They are all standard problems in visualization for the humanities—time/temporality, space/spatial relations, data analysis, and so on—but the investigation is posed in terms of innovative graphical systems informed by probabilistic critical hermeneutics. I carried on the preliminary work on each of these problems in specific research projects beginning in the early 2000s. These include two approaches to timelines: (1) temporal modeling (the system for creating analyses of the complex temporal relations in humanistic documents and corpora, experiential records, and other materials), and (2) heterochronologies (the modeling of systems of time-keeping according to specific and irreconcilable metrics, or comparative ontologies). The problems also include spatial modeling (spatial presentations created in accord with various factors, in a nonrepresentational approach to experiential, variable-metric mapping). Another problem is network inflection (modeling relations among elements in a standard edge-node formulation as node-edges that change across a set of variables that inflect the node-edge as a codependent entity). The final problem engages with argument structures and discourse fields (the shape of argument and its relation to evidence, combining automated techniques such as topic modeling with user-created argument structures); enunciation and the exposure of the hermeneutic lifecycle of data production (conventions for showing the modeling process of data and identifying subject positions within the apparatus of the interface). This chapter finishes with a final brief sketch of discovery tools as an additional interface into which modeling can be worked.

Intellectual foundations

The projects in this book, and formulations on which they are based, have developed over a period of two decades. Concerns about visual epistemology and its relation to textual and graphical forms have been integral to my work on the alphabet, visual poetics, book forms and formats, the graphic arts, and related fields for much longer.[8] I am not a philosopher by training. I am a scholar-practitioner educated in a succession of critical approaches. My intellectual trajectory merged structuralism, poststructuralism, critical aesthetics, theories of enunciation and subject formation in linguistics, film, psycho-analysis, and feminism with concepts of constructivist epistemology. This education occurred long before recent developments in what is termed *new materialism* had imagined its version of probabilistic encounters as a code-pendent relation between subject and object in a way that helps dissolve that binary. This combination of theoretical positions informs my work with deep convictions about *the primacy of visual knowledge within a probabilistic code-pendent framework rooted in humanistic interpretation (and critical hermeneutics).* These have not, as far as I know, been expounded elsewhere in relation to information visualizations, though elements of this approach can be found in cartography and arguments about relational space.[9]

The reader doesn't need to know this intellectual history, but it is help-ful, perhaps, to recognize that the work presented has had a long gestation period and was presented in multiple forms and many venues as articles, book chapters, talks, and funded research work in the last two decades. Pulling this together now, as the field of digital humanities continues to expand, but still without claiming the full cultural authority of critical interpretative work that is its core concern, makes my argument feel all the more urgent. Twenty years ago, the conceptual frameworks on which these projects are based seemed out of reach of implementation. Now, they seem ripe for realization and adoption. The obstacles to realization are not computational—they never were—but cultural. The resistance to the uncertainties and ambiguities of probabilistic hermeneutics introduces discomfort in many circumstances. But the desire for certainty of expres-sion, for visualizations that are strictly representational, has a psychological rather than intellectual foundation. The challenge is to shift that ground and make the force of argument, the rhetorical power of intellection, the heterogeneous structures of historical moments, cultural viewpoints, and individual judgment into a systematic and legible practice with appropri-ate tools and platforms. The timeline of these works also aligns with an increased interest in the decolonization of knowledge and all that this implies. This includes challenging assumptions long built into intellectual frameworks, such as those that privilege empirical methods and positivist frameworks. The situated conditions of knowledge production are politi-cally charged, not just intellectually inflected. Alternatives to rationalized approaches are endemic to critical thinking, particularly in a skeptical vein, and are essential to rethinking visualization in its interpretative role.

Another obstacle remains—the longstanding distrust of visual methods as primary modes of epistemological work. The humanities have long depended upon and been concerned with texts. In digital and computational activities, humanists have accepted the authority of quantitative and statistical approaches, perhaps because they appear explicit and unambiguous in their presentation of information. Almost paradoxically, the humanist, sure of theoretical, intellectual ground in the realm of discourse, seems to suspend that critical discussion when using the quantitative methods on which computation operates. The implication is that critical approaches belong to *content* in digital humanities projects, rather than to methods. But of course, this is wrongheaded and limiting. I have long argued that we need to insist on the use of humanities *methods* to develop computational tools and digital platforms.

And visuality? A long history of suspicion attends to the role of the visual in western thought. Philosophers beginning with Plato have distrusted visual images on a whole variety of grounds—as mere imitations, pale shadows, or much worse. To model interpretation using visual forms and with novel conventions pushes against many prejudices in the humanities in western culture. When we borrowed visuals from the natural and social sciences, the humanists seem to imply, at least we knew that we were borrowing systems grounded in representational strategies that correlated quantitative values and regular metrics. They were presumed to be solid on their own empirical terms. We borrowed them the way we might, unthinkingly, borrow a ruler—without asking when, by whom, or for what purpose the standard measures of inches and feet, yards, meters, or picas came into being and how they signify discursively—and visually. But the standard metric is an ideological construction, not a natural one. Also paradoxically, visual methods of its encoding render the ideological aspects nearly invisible—while also making it difficult to envision alternatives rooted in affective and interpretative activities.

To engage with a system of rhetorical expressions premised on the idea that visual arguments can become structured data, that informal and ambiguous graphics can give rise to formal expressions, may stretch credibility and patience. But that is the task undertaken here, along with the justification for the effort, because the cultural authority of critical positions cannot be asserted, enacted, or demonstrated without adequate means of expression in graphical ones. The politics of knowledge production are intimately bound to the methods through which it is enacted.

1 Visual Knowledge (or *Graphesis*): Is Drawing as Powerful as Computation?

We are not necessarily accustomed to thinking about visual images from a theoretical perspective rooted in problems and questions of knowledge. And yet, visual images have capacities for production and presentation of knowledge that are unique and also fully integrated into common understanding and daily activity across a wide range of disciplines. Let's begin by addressing the status of visual epistemology directly and then return to problems of digital work and humanistic interpretation.

Visual expressions serve not merely as representations of existing knowledge, but as primary modes of knowledge production. They have the capacity to produce and embody information, not just represent it. This assertion suggests that experience, feeling, sentiment, conceptual and intellectual schemes can be expressed as visual statements (in a declarative manner). But it also asserts the capacity to make claims (propositional statements) or provocations that may be tested, justified, or argued using visual evidence. Visual epistemology, therefore, can be both declarative/descriptive and propositional/provocative. Demonstrating that this is the case is the task here, and showing the ways that visual images perform their epistemological work is central to the argument.

Before beginning, a general statement should be made to clarify the distinction between representational and nonrepresentational approaches to visual forms of knowledge, since the contrast may be unfamiliar. Representational forms assume a secondary status; they are surrogates, and stand for a preexisting, a priori, already formulated knowledge in the form of a graphic statement, notation, or visual phenomenon of some kind. A map may represent a territory (even though it contains the distortions of projection systems and of course many features that never show up in the physical environment—approaching a state capital on a highway one does not see a giant star in the landscape, for instance). Similarly, a portrait may represent a specific person, a graph may represent a data set, and an anatomical drawing may represent a body or its systems and parts realistically or schematically. These are common representations. They are not equivalents. They are not identical to what they represent. They are surrogates that exist in various degrees of remove and transformation from what they reference. They translate knowledge into visual form.

A nonrepresentational visual expression creates information or knowledge in a primary mode. An architectural sketch *brings forth* the image of a building, a geometrical diagram *creates* a proof, a drawing *produces* a form hitherto unknown, an act of connecting one or more words in a text with a line *creates* an interpretation, or a drawing of an arrow *creates* a model of time or temporality. In these examples, the production of the visual image produces something new, it does not reproduce something preexisting. Graphical calculus, in a precomputational era, was a means of arriving at results, not a method of simply displaying them.

This distinction matters because while the representational role of visual images is well known and understood, the nonrepresentational aspect is less familiar. Representational images have a relation to an existing referent (something to which it refers). But the nonrepresentational status of visual images (which seems counterintuitive at first as a notion) is harder to grasp. Images always seem to suggest a relation to an existing referent (they *seem* premised on resemblance). But does a visual expression always have to be dependent on a preexisting referent? At stake in answering this question is the claim that graphical work *is* intellectual work, as a primary mode—a claim supported by practitioners across a wide array of fields from theoretical physics to engineering and design, as well as those in the conceptual and visual arts. If we modify this argument one step further, with the constraint that the main interest here is to address ways in which humanistic approaches to visualization express an interpretative approach to knowledge, then our scope narrows. But support for the experimental work follows. Before we can engage in detail with these particular and focused concerns, a more general view of visual epistemology needs to be put into place.

Visual epistemology

The range of disciplines that make use of visual approaches to epistemology is broad, and issues of style should not be allowed to confuse the basic functions of images. This remains true at the specific intersection of digital methods and visual forms of knowledge production and interpretation. The capacity of images to present knowledge is not a matter of form, but of the relation of images to referents: knowledge can preexist and be shown or be created by the image.

Graphical forms include any visual arrangement of marks or visual forms on a support surface or substrate (paint on canvas, pen on paper, pixels on a screen). These forms are literal and visible and need not be put at the service of pictorial illusion. All graphical artifacts are comprised of marks or traces organized on a surface (broadly construed to include a wide range of materials, projections, immersive environments, and so on). They embody knowledge through the combination of symbolic codes and structured relations of these elements in a field. Individual visual instances (a graph, a drawing) can change over time, either through accretion or deletion. But methods of making graphics also change with conceptual and technological innovations, such as occurred with the invention of animation.

Graphical knowledge cannot be grasped in any self-evident way. For one thing, images have no simple correlation to language—no single word can be used as the equivalent of a single line, even if language can be used to describe it. On its own, a line or mark can be highly ambiguous—highly specific and infinitely unique. Like any other form of human expression—natural or formal language, mathematical notation, bodily gestures, or other signs and codes—visual imagery becomes more stable and more useful when interpreted in combination with a linguistic gloss or statistical base. This statement doesn't negate the specific properties of visual forms—many of which have no equivalent in language or formal systems. But it should put to rest any lingering naïve formalism premised on the idea that visual representations communicate directly and simply. Images are not self-evident. We learn to interpret them through learned approaches to encoded expressions that provoke a response for cognitive processing. In other words, reading images is an acquired skill embedded in cultural and historical circumstances.

Visual and digital

Visual forms of knowledge production do not depend upon digital techniques. We use visual skills to distinguish faces, read graphic novels, and navigate a complicated route without any digital technology. But computational capabilities offer unique opportunities to make use of visual methods. Not only are screens and devices organized with graphical interfaces, but the processing of information in digital form makes it easy to create visual displays for all kinds of purposes. Conventions for graphical display have become familiar, legible, easy to read—and equally easy to take for granted. We hardly see the workings of visualizations and take their forms for mere statements, representations of information with which we quibble for their accuracy or effectiveness, while ignoring the larger issues of knowledge claims embedded in their use.

But received conventions can be extended to model generative and/or interpretative, as well as representational, approaches. Much is contained in that statement, and it will need to be carefully expanded. Even the most basic aspect of my argument—that visual methods can *produce* knowledge and *enact* interpretation, not just serve as representations or displays—is considered polemical by many.[1] To insist, further, that interpretative methods have their own specific requirements and possibilities, that these are rooted in probabilistic hermeneutics and critical theory specific to the humanities, only multiplies the challenges. In this first section, I will address the basic issues of visual knowledge in digital environments, leaving the discussion of probabilistic approaches for the next chapter and the nonrepresentational modeling approach for the one after.

A political agenda inheres in this project, which is to demonstrate that value-laden (ideological) operations of knowledge production can be exposed through methods that attend to the historically situated, culturally located, and individually inflected systems (described above as *systems of enunciation*) within which they work. Attending to these ideological operations is not the exclusive responsibility of humanists, but I would argue that humanists possess methodological tools that are particularly well suited to this task. The connection between these tools and the terms on which visual expressions function in digital work is direct. Ideology (cultural value) appears in every graphic, layout, format, bit of iconography (as well as in interface and navigational features to be addressed in a later section), even as it disappears through the familiarity of conventions.[2]

The first challenge is to address the question of whether images produce knowledge, and if so, how and what kind? Is there a specific *visual*
epistemology? Following this, we can ask how engagements with modeling
interpretation might extend or challenge graphic conventions used in digital humanities. The specificity of visual epistemology and the particulars
of graphical means feature in this argument, but these are concepts that
do not belong specifically to the digital realm.[3] The principles of *graphesis*
as the basic concept of visual forms of knowledge production will be dealt
with more summarily here, while the relation between visuality, interpretative methods, and digital humanities will be brought to the fore.

Visuality, perception, and cognition

Human cognitive and perceptual capacities make use of vision for a range
of complex activities that have nothing to do with images. The literature
on vision and the brain is supported by empirical experiments that explore
these capacities in a scientific way.[4] The historicity of vision and cultural
conditions of perception also factor into these studies.[5] But the connection
between vision and knowledge explored here is not grounded in perception as a physiological fact. Instead, it is focused on the ways that specific
properties of graphical inscription and notation can be contrasted with
those of binary code in terms of their inherent qualities as well as critical
approaches. In other words, my attention is on the knowledge-producing
(epistemological) properties of graphical images in digital environments,
rather than on the mechanical study of eye-tracking or click counts and
other studies of visual processing.

As already noted, a vast inventory testifies to the long-standing relation between knowledge and images. This includes scientific illustration,
maps, charts, graphs and diagrams of all kinds—from those charting blood
flow and plant morphology to ones used to track the movement of stars in
the heavens and orient sailors toward their destinations.[6] It also includes
account books, civic monuments, architectural plans, and city signage as
evidence of knowledge structured visually and organized graphically. The
list could proliferate endlessly. Most would agree that these images serve
to represent knowledge within a wide range of disciplinary domains, but
precisely how they are themselves *arguments* about knowledge or belief,
or are able to *produce interpretation* is less clear. Current work on the way
scientific illustration and natural history are bound together has advanced

these insights, but the underlying assumptions of this line of scholarship privilege scientific understanding as a legitimating factor in assessing the images.[7] The value of the images rests on their capacity to communicate information in accord with supposedly neutral assessments of "accuracy."

The long-standing realization that direct observation is insufficient to ensure accurate perception or depiction can be amply illustrated. Whether we look at the peculiar historical drawings made from beached whales that provide them with fanciful fangs, or see the detailed male genitalia observed on mandrake roots, or wonder at our inability to see a figure in the gorilla suit walking through a group of basketball players in a viral video, we see the limits of assuming any direct encounter between perception and cognition.[8] No "innocent eye," exists, as the twentieth-century art historian and psychologist Ernst Gombrich's characterization of this mythic notion made clear.[9] We see what we know and what we expect to see as much as we process phenomena according to their visible features. Theories of vision go further than this mere qualification of direct observation into psychoanalytical, anthropological, sociological, and ecological realms. For instance, James J. Gibson, with his theory of ecological vision, describes the combination of selective and adaptive characteristics that are combined in visual processing.[10] Not only do we learn to see in accord with our species-specific needs, but individuals also "learn" to see differently. Further experimentation notes that the eye itself changes, as does the full visual system, when exposed to particular stimuli.[11] The codependent nature of the perceptual and cognitive psycho-physiological system shaped by behavior and experience, as well as psychological patterns and dispositions, is further complicated by the integrative work of the cerebral cortex, which processes several sensory inputs in its activity in a process referred to as multisensory integration. So, not only do graphical modes have the capacity to produce interpretative models of experience, the visual sensorium is itself an interpretative system and one that is modeled by the multifaceted features of perception and cognition across sensory inputs.

Within the discussion of scientific images, intellectual value is always assessed with a single criterion: how well do works of artists in a particular period conform to or reflect contemporary scientific attitudes toward knowledge? Questions of aesthetics factor into the discussion only with respect to the task the images are presumed to serve. Take Claudia Swan's discussion of color in drawings of plant species, for instance. As she reports in a study of botanical illustrations, eighteenth-century scientists chided the artists who were drawing renderings of specimens for creating misinformation through their use of (observed) color variation. The color created the impression that these drawings were records of multiple species when they were clearly merely individual instances of a single species with shared morphological structure.[12] Swan's discussion does not challenge the botanists' claims to cultural authority, working as it does within empirical approaches. She is an art historian, not a scholar of science and technology. She does not question the empirical accuracy accorded to scientific knowledge. In her view, epistemological authority resides unquestionably in the correlation of image and object on terms determined by scientists. They assume that an object can be perceived objectively, independent of a user's own historical disposition. This assumption permeates the historiography of work on images in the service of botanical, medical, and zoological knowledge across the realms of "natural" history. The question of what other epistemologies might be available in these historical and cultural moments is never asked. What other models of knowledge and visuality might be considered besides those of empirical science? How might this change with a consideration of phenomenological approaches, for instance, or emotional or symbolic ones? Is the ability to read the properties of a particular pressed-flower specimen (the moment of its being found and preserved) or discern a particular kind of damage (sign of a large animal or of a mower) a form of knowledge? This is a question about the status of knowledge, not merely its visual expression. In the empirical sciences, the *only* form of knowledge that matters is that which is objective, user-independent. The attitude is premised on belief in perception as direct and uninflected by cultural conditions. The claim is that empirical knowledge operates authoritatively without historical or cultural assumptions or influence. Such an attitude serves a purpose within certain domains, but does not exhaust the capacities of images to generate knowledge. Visual epistemology has long been subject to these powerful belief systems, seriously limiting discussion of properties of visual rhetoric and inscriptional specificity (the unique properties of each instance of graphic work). The empirical tradition ignores (and delegitimizes) knowledge grounded in nonempirical methods, even though textual and cultural studies have long ago exposed the limitations of empirical claims to universal truth—and the biases and blind spots on which they operate.

Images of nonvisible phenomena

Many aspects of natural, social, or cultural phenomena have no visual corollary. Even in the physical sciences, recording the work of forces, vectors, atmospheric and thermal conditions, and so forth is dependent on various conventions, not observation. For instance, contour lines that show altitude on maps, and wind arrows, isobars, and high- and low-pressure systems are all presented in graphical codes, but that does not make them "real" parts of the natural system.[13] Peter Galison has traced tensions between two research traditions in the sciences that make use of images—pictorial presentation and computational production—as dual parts of an epistemic practice that structured much of the experimental activity of modern physics.[14] The image of the atom as a small solar system of particles in orbit, for instance, is not based on observation, but on an idea of how to represent a model of a phenomenon. The creation of images, such as traces in a cloud chamber, may affirm or undermine a theoretical hypothesis. But conceptualization of phenomena is often as strongly influenced by the models made as by observation (think of gender categories as an example where changing models have changed perception). Lab work may be done visually or it may use visual methods just to show results (such as the display of changes in temperature across time). This is true in the humanities and social sciences, and the role of visualizations in reification of constructs and concepts is powerful. Even very basic features, like the separation of temporal intervals into years or days carries assumptions that may or may not accurately reflect the phenomenon being represented in these graphical-conceptual structures. Our understanding of forms of knowledge is often only a knowledge of forms (or, to put it another way, *how* we know often shapes *what* we know). Recent experiments in creating novel structures in organic chemistry, or biological organisms through visualization techniques, suggest that the manipulation of visual models to generate new structures is making aesthetic research a vital part of scientific innovation. Properties and behaviors get defined as an effect, not an a priori given, in ways that have been a part of architecture and applied design for centuries.

Mathematics is full of graphical forms that do meaningful work, such as matrices or formulae. These support high-level operations and manipulations. As already noted, visual conventions for such familiar procedures as adding or subtracting numbers, or using decimal points to define place-specific values are taken almost for granted though they perform essential functions. Formal structures like Venn diagrams provide dramatic examples of expressions that *do* the work they express; they do not simply represent it. And in the specialized field of visual calculus (mentioned above), graphical methods were used to find solutions to complex problems before computers were capable of automating these procedures.[15] The rationalization of surface as a ground on which to work (by the use of grids, or metrics, or values on an *x* or *y* axis), and which provides a stable space for graphical presentation, is so fundamental we forget to regard it as a conceptual act that advanced knowledge. Similarly, the creation of ground lines for production of straight lines of glyphs, cuneiform marks, or early writing, established conventions of linearity that still prevail.[16] Likewise, the hierarchy of top and bottom, the dominance of left-to-right reading in much of western culture, and other structuring principles of graphical systems are embedded in the visualizations we use almost without thinking.

In addition to contemplating the semantics of format and arrangement, we can consider the basic forensic evidence of graphics, the indexical traces of events or activities left behind as lines or marks. These marks provide evidence, ways to read back into events and to reconstruct them. The trace of a signature, the imprint of a seal, or the scuff marks on a section of wall or floor are all indexical signs. Anthropologists read the markings on walls, the imprints of hands, feet, and bodies in hollows and seats around a site, according to their graphical properties to discern meaning and value. The marks of editing, reworking, and rewriting that accumulate on a manuscript are also read as graphical traces of past events. These require systematic decoding and analysis in other systems of description that are in turn organized with graphical means for critical editing and bibliographical work. The information embodied in formatting features contains codes for reading, instructions on how to interpret each word or line depending on its placement. We rarely pause and consider these codes—or the multiple ways in which our reading of any text or image is governed by the site in which it appears. Formal, official, domestic, educational, public, bureaucratic frameworks—we read these conditions in moving from context to text and back again.

But, as already hinted, graphical form can also be a semantic, original, and primary means of generating knowledge. Consider the task of creating a plan for a building, sketching the design of a traffic system or a workflow diagram. These methods bring into being something that has not existed before. The image contains information that cannot be replicated very readily in textual or numerical form. If I want to chart a path or create an instruction sheet for dance steps, verbal description is best as a supplement, not a primary mode of knowledge production. An implied question is embodied in every napkin sketch of an imagined structure or product outline: "Will this work?" In essence, each sketch is a visual thesis, a proposition about possibilities of knowledge that would have to be tested with evidence to see if it is a justified belief.

All of these practices—scientific illustrations, traces, encoding, experimentation—are part of the inventory of methods of graphical epistemology.

Aesthetic images as knowledge arguments

In addition to images from science, engineering, or other realms where they serve applied tasks, another inventory of visual knowledge is contained in the history of artworks. These images are rarely approached as knowledge objects, but instead, are studied for their style, form, provenance, social impact and influence, content, iconography, theme, treatment, composition, authorship, ideology, politics, influence, forgery and a host of other considerations. These issues eclipse attention to epistemological claims when dealing with art historical and aesthetic works. In fact, the very question of whether an image makes an argument or embodies justified claims of belief occupies only a small number of art historians and theorists. And yet, it can be argued that every visual image makes an implicit or explicit argument about visual knowledge and the terms on which the relation between knowledge and its production and presentation are understood.[17] A single example will serve to make the general case.

The example focuses on two images by Rembrandt—a *Self-Portrait* of the still-young artist (1627) and *The Anatomy Lesson of Dr. Nicolaes Tulp* (1632).[18] They are relatively close in time to each other, especially given Rembrandt's long career. But they are stylistically distinct in deliberate and significant ways that, I suggest, have to do with their epistemological assumptions. In the self-portrait, the shadowed eyes, blurred focus, and obscured brow of the young Rembrandt pose the question of whether or not—and *how*—one might know and picture one's own identity, one's "self," in any objective way.[19] The confusions of self-perception are posed visually, as questions of knowledge, in ways that cannot be easily resolved. What can be seen? How is perception able to function independent of the confusions that cloud the visual apparatus with subjective judgments? In the early self-portrait, the artist is so far within his own self-perception that he focuses on *that* as the very theme of the work. He makes a vivid argument for *observer-dependent knowledge*, the very condition of being within the interpretative process, by giving visual expression to this question: Can a subject know himself or herself as an object?

That absence of resolution with respect to the relation between perception, knowledge, depiction, and identity of the subject as object is in direct contrast to the detailed and nearly surgical inscription of the individual portraits of each of the figures in the *Anatomy Lesson*. If the self-portrait is a study in modeling a subjective position, the *Anatomy Lesson* is a demonstration of belief in empirical approaches to knowledge. With evident faith in the capacity of vision to be as precise in its relation to an object as the surgical instruments that are laying bare the sinews and muscles of the arm for clear examination, Rembrandt shows us the identities of the medical figures through depiction of their features. They are knowable by their physiognomies, and these can be represented, objectified, made into images. The terms of knowing and recognition are visual, as is the anatomical knowledge offered by the depiction of the corpse and, presumably, the book propped open at its feet and the document held in hand by one of the observing physicians. Forms of visual knowledge and commentary upon their workings into and as images are multiplied in these two works, and the contrast could support a much longer discussion.

But the principle at work here—that images embody assumptions about knowledge—can be broadly applied. All images embody assumptions about visual epistemology—its limits and capacities, and its specificity. A study of an early modern work by Paul Cézanne could call forth a different but equally pointed set of observations about the assumptions of knowledge on which it is based, as could a satirical work by William Hogarth, a portrait by Alice Neel, or a conceptual work by Robert Gober. Cézanne asks how we can know the volumes of air, space, and objects as components of the world and find a language adequate to express them. Neel suggests that all sitters reveal the contradictions of what is apparent and what is concealed by appearances. Hogarth shows that knowledge is circumscribed by convention: the patterns and orders we perceive may be as deceptive through their framing as the position of objects in his 1754 satire on visual perspective. Or pretensions and affectations may be concealed by the codes of dress and decorum of the figures whose worlds he depicts. Gober shows how vision encodes cultural bias, and then surprises us with alternatives that show the depth of these entanglements.

For Rembrandt, degrees of focus and precision of delineation are visual methods in the service of his argument about differences in types of knowledge, those of self-observation and observing subject. His argument is made in his *methods*, not only the composition or iconography he uses. The two images offer their crucial contrast graphically—between the contingencies of a situated, hermeneutic knowledge and that of a presumed objectivity, (fictively) independent of an observer. He differentiates these positions through the vague blurring of one and the tight finish of the other. The self-portrait is a demonstration of the way the conditions of user-dependent interpretation can be exhibited, called to attention as situated knowledge that cannot come into focus because there is no "outside" place of observation. The self-portrait is always inside of the hermeneutic process in ways that are inescapable.[20] The portraits in *Anatomy Lesson* demonstrate the presumption of observer-independent knowledge, as if these figures would have the same appearance under any circumstances and in the gaze of any observer—and as if identity were isomorphic to physiognomy carefully observed.

Not every image poses explicit questions about observer-dependent and observer-independent approaches to knowledge. But every image (as suggested in the cursory sketch above) is premised on one or the other of these positions and accompanying issues of the construction of vision, its cultural features, historicity, and psychological aspects. If the long history of connections between seeing and knowing can be teased from any and every instance of visual expression, then how does the use of digital imagery push these concerns? How does the difference between visual forms and expressions used *to model interpretation* and those used *to visualize information* become formulated within a discussion of long-standing concepts of *mathesis* in relation to those of *graphesis* specifically with respect to digital environments? The term *mathesis* describes a system that represents knowledge with the same formal explicitness and lack of ambiguity as mathematical notation. Its symbols are repeatable and always have the same value. By contrast, *graphesis* describes a system in which every instantiation is specific, characterized (however minutely) by individual differences.

Mathesis and *graphesis*

In the late 1970s and early 1980s, the proliferation of digital platforms was accompanied by a flurry of theoretical characterizations of code as "pure difference."[21] This idea of a "pure" difference, one that was constituted without material instantiation, was completely false, but it found many eager advocates nonetheless. A handful of citations suffices to show the stubborn persistence of these notions. For example, take Garfield Benjamin's statement: "Binary logic is built upon the purely formal numbers one and zero, exemplifying Deleuze's pure difference and Žižek's minimal difference, functioning as the coalescence of meaning with the real in the fundamental positing of a digital universe, virtual-existence then appears as a realm of the imaginary in the superficial light of the interface screen."[22] Or, Michael Eldred's: "Basically, an ordered sequence of zeroes and ones (nothing and something, pure difference) is transmitted which at the other, recipient's end can be and must be recomposed in such a way that the appropriate result … is brought about."[23] And Alain Badiou's: "It so happens that the difference between zero and one is very important in our digital world. The basic difference between zero and one is the origin of all differences, it is pure difference, the paradigm of all differences, as we see with the concrete example of the development of binary code."[24]

The mythology and misunderstanding in these statements expose a longstanding philosophical prejudice against materiality (not only in its graphical form), as if the horrors of embodiment could be fully avoided at last. The "purity" of code, these approaches implied, inscribes ideas in an "immaterial" form. This pure form, mathematical—metaphysical—in its logical structure, could triumph over the base stuff of physical embodiment. We might imagine that the awful dross of the material world, with its bodily associations, could at last be discarded and disregarded. Ignorance of the complicated materiality of digital technology underpins this belief, but that is of less importance than the impulse to embrace an abstraction rooted in a value system of purity associated with distance from material instantiation.

In western thought, images have long been associated with idolatry and temptations of the flesh. We have only to conjure the image of the Golden Calf from the Old Testament to see how images are associated with practices that turn a people toward adoration of an idol and away from worship of an ineffable deity. Platonic hierarchies, later absorbed into Judeo-Christian ones, consider images a debased version of ideas which alone can approach truth. When truth is understood as universal, and faith is premised on the ineffable, these tenets of belief hold as foundations of knowledge as well as other forms of belief. The idea that knowledge might be specific, local, contingent, instantiated, and associated with sense perception and cognition is at odds with these traditions and their highly influential premises. Images are stigmatized by this kind of theological and philosophical baggage. This led to assumptions that digital images, because of their relation to "immaterial" code, might not be. Here we see the persistence of rhetoric to affirm the new "purity" of code as "difference." The fact that this "difference" had to be embodied in complex layers of silicon, software, platforms, devices, displays and so on was conveniently overlooked.

Perhaps more important than the characterization of code as "pure difference" is the contrast we can make between *mathesis* as basic binary code (formal, mathematical, logic meant to serve to inscribe all knowledge in a single, unambiguous, logical form) and *graphesis* as an inscription that either produces or represents information and/or knowledge in visual form through those properties of variety and specificity mentioned at the outset. Significantly, the investigation of *graphesis* will provide a way to return materiality to code, rather than the contrary, while also distinguishing it from *mathesis* according to the criteria of formal structure, ontological identity, and performative capacity.

Mathesis, as already noted, is defined as "knowledge represented in mathematical form, with the assumption that it is an unambiguous representation of thought."[25] It embodies the attempt to engage a formal system of knowledge production that follows the laws of mathematical logic applied across many other domains of knowledge production.[26] The goal of *mathesis* was disambiguation, clarity, and precision of expression and combination. Unlike the messy reality of natural language, a completely formal language would make use of all of the properties of logic and mathematics to compute ideas, arrive at irrefutable conclusions, even calculate outcomes of moral and theological debates. This quest for the "laws of thought" through this formalization of knowledge has a long history, but the term *mathesis universalis* is associated with Gottfried Leibniz and René Descartes.[27] Their quest for a universal mathematical system also coincided historically with a political desire to communicate across language groups and cultures. *Mathesis* was supported by the utopian dream of utterly clear and universal communication.

These quests gave rise to several of the universal language schemes and innovative sign systems that proliferated in the seventeenth and early eighteenth centuries as philosophical experiments. A handful, like the extensive project of Bishop John Wilkins, were designed so that they embodied the structure and organization of classification systems as a kind of visual code.[28] In such a code, the category of plants could be indicated by an upright stroke, and animals by a horizontal one. Details of the place of any specimen within the hierarchy of other members of its kingdom was indicated by other strokes, dots, and features. The aspiration was for a "word's" value or meaning to be deciphered directly, by decoding its place in this system. Each category—animate or not, animal or vegetable, vertebrate or invertebrate and so on—would be encoded in legible signs.[29] The many fallacies and assumptions in this approach include a conviction that language mainly represents things and that ideas are universal, cross-cultural, and objectifiable within stable and discrete logical systems. This quest—for signs that could communicate directly to the eye—was also partially inspired by a misunderstanding of the pictorial qualities of hieroglyphics as well as Chinese characters, both of which were taken to represent things unambiguously.[30] These popular misunderstandings were widespread in the late Renaissance, and persisted into twentieth-century work by, for instance, Ernest Fenollosa, the scholar who had such a strong influence on poet Ezra Pound's idea of the ideogram.[31] In the case of Leibniz's system, the goal was to create a formal system to calculate thought (or in which thought could be computable), not just represent things. His interest in the power of the *I Ching* as a combinatoric system was an inspiration for this approach. The quest for a universal, visual character and for a computable formal system of communication shared certain convictions about the character of mathematical notation and its logical systematicity.

This quest to discover an analogy between logic and thought persisted well into the twentieth century. Some projects were inspired by nineteenth-century figures like George Boole. His 1854 work *An Investigation of the Laws of Thought* is a precursor to the logical positivism that informed much of Anglo-American philosophy, Vienna School studies of language and sign systems, and models of computer intelligence.[32] These philosophers did not confuse thought, a mental process, with formal logic as a representational system, but in popular perception this became the case as computers became widespread. The work of many early researchers in artificial intelligence, such as Norbert Wiener, was premised on a belief that the human brain could be compared to a computer and that a formal system might be developed to serve as a corrective to the faultiness of natural language with its embedded dependence on circumstances and nuances of communication.[33]

These multiple strains of intellectual inquiry all have a place within the larger framework of *mathesis* understood broadly as the belief that the authority of logical, formal structures inheres in their capacity to both create and represent knowledge in expressions that have no ambiguity to them, that are computable, and that are completely independent of their circumstances of use. The brain-computer comparison became the object of criticism from a range of philosophical viewpoints, famously, those of John Searle and Hubert Dreyfus.[34] Searle's "Can Computers Think?" was an argument against artificial intelligence based on a distinction between mechanistic processing and human capacities for intellectual thought.[35] Dreyfus's *What Computers Can't Do* (1972) was premised on the distinction between symbol manipulation in formal systems and unconscious intuition as well as embodied and lived experience as a foundation for human thought. His landmark work has featured in many debates that still swirl around comparisons between human brains and computational processes.

Now we can turn to a critique of *mathesis* through a contrast with *graphesis* in terms of modes of expression, as well differences in attitudes about the foundations of knowledge in its creation, communication, and use (rather than in historical and philosophical misunderstandings about language). *Mathesis* is aligned with systems that privilege disambiguation, stable notation, and logical structures, as well as universal approaches to standard notation systems. *Graphesis* allows for ambiguity, instability, and rhetorical expressions and privileges the specificity of inscription. Graphical marks do not rely on difference for their value, but on specificity—the fact of each line or mark's unique identity and quality. Even though they may be specified by algorithms, graphical lines are still "drawn" in digital environments. The implications of the differences between notational (mathematical) and inscriptional (graphical) practices (see below) leads to consideration of the underlying data structures and relations between these and graphical forms as well as of the visual expressions themselves.

Because digital technology operates in ways that are discrete and unambiguous at the level of the encoding and processing, it has often seemed to be the fulfillment of the quest for *mathesis*. However, all symbolic systems, whether analogue or digital, can be put at the service of ambiguity and complexity at a higher level of representation. We can write nonsensical statements in a formal system quite easily, so long as the syntax of the formal expression conforms to the rules. We can use a line to draw a shape that defies meaning, even if it is stored unambiguously in an algorithm.

The distinction between *mathesis* and *graphesis* is not the same as the distinction between digital systems and analogue systems. The concept of *graphesis* draws on many principles of analogue forms of knowledge production which apply to digital inscriptions as well. One of the challenges for humanists wishing to create visualizations in digital environments that are adequate to the rhetorical and epistemological requirements of interpretation is to demonstrate the ability of computational systems to inscribe the specificity and particularity that are inherent features of visual knowledge production. Therefore, these features need to be identified and described and their role in knowledge production understood.

Graphical specificity

In his 1953 landmark study *Prints and Visual Communication*, William Ivins showed that engravings made possible the production of "exactly repeatable statements" in visual formats. He felt this made a contribution to the advancement of modern science.[36] Beginning in the fifteenth century in Europe (and considerably earlier in Asia), printed images in scientific and technical publications created stable, repeatable, shared references, initially through woodcuts. The advantages of printed images were twofold. They could be circulated to create common understanding and visual reference, and they had particular graphical properties imposed by the technologies of the media in which the lines were produced and reproduced from wood and then metal. These supported the production of graphic information in what Nelson Goodman, in *Languages of Art* (1968), identified as either *allographic* or *autographic* modes (on which more below).[37] The increased detail of line and tone afforded by copper and steel plates extended the graphic range—and consequently the authority—of images. Greater detail meant more information, and more information meant more accuracy and verisimilitude—in the same way, higher-resolution file formats are informationally richer than lower ones (though of course not always more accurate). Prints served to mediate scientific knowledge in every sense. They were media, a means of inscription that embodied and communicated information in graphical codes. And they functioned to mediate—to serve as a site for focused, intersubjective exchange among professionals—contributing to the creation of a scientific community and consensus about the natural world and knowledge of it.

Ivins didn't assume that a particular method of making (or technology of production) produced a particular response. He was not deterministic, but he did provide a way to read visual effects of media. The trained eye can read image structure, but also production history, by distinguishing different kinds of marks, lines, tonal values, color ranges, densities, textures, and patterns. The surface of a currency bill, for instance, with its raised ink lines left from an intaglio plate, has a graphical-tactile character that is essential to its operation (even in an era of embedded holograms and specially treated fibers). Visual translation from one medium to another always reveals the effects of transmission, the graphical code that literally inscribes knowledge. Ivins's passionate dislike of the ultrarational technique of steel engraving was a dramatic instance of his attention to the affective power of image technology. He found the mechanistic dot and line patterns of the engraving roulette tool antithetical to aesthetic expression. Extreme cases always show the rule—the softening effects of paint on black velvet can render the most rigorous image kitsch, and the use of a medium can create cognitive dissonance simply by force of the rendering. An inventory of the various properties of graphical media does not supply a vocabulary of semantic values any more than dreams can be reduced to an inventory of symbols or styles in art to a particular set of brushstrokes, pigments, or even compositional methods. But in reading images as part of knowledge systems, Ivins's approach offers a fundamental skill.[38] We can read file formats with similar approaches in understanding what kinds of visual information they can record and how, and how in storing graphics in various ways, they embody models of epistemological understanding. The creation of an image as a vector graphic, a line drawing, a pixel tapestry, an RGB or a CMYK file has implications for precisely what information is constituted in its inscriptional trace. These technical and mechanical aspects of images provide a foundation on which the distinction between inscriptional (continuous value) and notational (discrete value) methods make sense—and on which the substantive differences in their functionality can be assessed.

Notation and inscription

René Thom, the mathematician and philosopher, once wrote that "for a phenomenon to be an object of science, counted in the common (and, in principle, eternal) heritage of scientific knowledge, it is first necessary to describe it."[39] Then he noted that only two such descriptive techniques exist—"natural language and mathematical formalism." He conspicuously excluded graphical description.[40] Thom must have been aware of the substantial role visual images have played in producing and communicating knowledge in the history of the natural and physical sciences, but he considered graphical formats inadequate for accurate knowledge production and transmission.

But as we have seen, even before the existence of print technology, visual images served varied functions in connection with knowledge—from the representation of information in condensed, legible form, to the expression of complex states of mind and experience. Maps, graphs, diagrams, illustrations, pictorial images of all kinds, even handwriting and inscriptions, provide information through graphical means (as images) but also through their specific visual features (texture, syntax, color and other characteristics). Thom's oversight, if it was an oversight, or decision, if it was deliberate, thus seems peculiar for so astute a thinker. His attitude is reminiscent of the denigration of vision in modern critical theory described by Martin Jay in *Downcast Eyes* (1993).[41] As explanation, we can again invoke the long tradition of logocentrism that dominates western epistemology, even though, paradoxically, sight was always privileged in the hierarchy of the senses. From antiquity onward, the eye's capability was associated with reason and enlightenment in ways that taste, touch, smell, and even hearing never were. Vision allowed perception at a distance, without direct contact or fleshly stimulation. But the promiscuity of images, their capacity to serve so many functions (entertainment, pleasurable distraction, communication), calls their knowledge bearing (epistemological) and moral status into question. Even recent critical studies admit that visual images are a less stable form of knowledge production than language. Large philosophical issues and deep-seated prejudices haunt us at every turn.

Something substantial had to have been underlying Thom's pronouncement. Perhaps the main source of discomfort about visual images for any formal logician is that graphical signs are not stable in relation to their referents. Where a number can signify a quantity without ambiguity, an image is open to association and interpretation. Even showing the concept of "two" in any graphical form, beside that of a number, creates suggestions and association. Two pies, two apples, two kittens—the "twoness" of these is difficult to separate from the other features of the images. Once the concept of "two" is extracted from the objects, however, it can be represented without difficulty with a numerical symbol. But with images, the problem extends to the level of the marks, the notation system. The basic code of numeric and alphabetic notation (and/or other written characters, such as Chinese, Japanese, or chemical symbols) is fixed and finite. Every letter or character can be distinguished unambiguously. Their combinations can be read, but more importantly, the components of written language can be disambiguated. Alphabetic signs do not have a comfortably fixed relation to morphemes or phonemes, and the same letters are used for a variety of languages, but they can be distinguished from each other. Their disambiguation at the level of signs is not a problem, even if their connection to various references may be complex and sometimes ambiguous.

Numerical codes are even more stable than linguistic ones, and at any level of combination or complexity, a numerical value and its sign remain in a stable relation. By contrast, a word and its meaning do not remain fixed, and the study of language has long concerned itself with the slippage of word and meaning, text and referent, value and use. But no equivalent to either alphabetic or numeric code exists in images. A line can be short or long, a dot can be round or misshapen. A stroke can be wide or thin, vary in width and weight. Where do the basic units of code begin and end? Of what does the basic visual system consist? Again, the answers to this question have generated a considerable literature.[42] But no amount of analysis can resolve the unresolvable problem that, unlike language, images do not have a stable basic code.

Thom was right, therefore, to be anxious about graphicality, since it could not be relied on to be consistent in its execution (as a fixed set of notational signs) or its representation (in any stable relation to its references). A shift in size or weight could remake the value of a line and then change its meaning within an image in ways that could not be accounted for in any logical system. The resistance of graphicality to systematicity is one of its fundamental (epistemological) properties, but for Thom it was a fatal liability.

Because of his empirical disposition, Thom was concerned with notational systems, not inscriptional ones. The distinction is important for the arguments about the relation between visuality and epistemology. Both the contrast of *graphesis* and *mathesis* and the distinction between notation and inscription are asymmetric, but are crucial to the discussion of the specific quality of graphical production and presentation of knowledge.

Goodman's work, alluded to above, is a still-unsurpassed systematic study of visuality framed through a formal, analytic approach. Goodman's definitions and analyses still hold, even if his focus did not address any of the dimensions that historical and ideological study bring to the understanding of images. His work was conceived under the influence of a modernist desire to create a complete understanding of "the language of form," as if images were, like mathematics, subject to universal laws that worked outside of their temporal and cultural location.[43] Goodman's work still provides useful definitions and distinctions, such as the one he makes between autographic and allographic notation systems. The term autographic refers to those modes of inscription that cannot be transcribed without information loss. So, a signature is comprised of the particular loops and curves and muscle patterns that comprise it. Translating handwriting into a typewritten name or typographic rendering would take away vital information and authority that inheres in the signature (though digital signatures have eroded that particular aspect of functional distinction). Allographic systems, on the other hand, can be transcribed, remediated, and copied in another modality, or so Goodman believes, without losing their informational substance. They are notational rather than inscriptional. How far exactly the defense of allographic systems as invulnerable to information loss would go is worth considering. Take the example of a typewriter as an allographic system. Change its font into a typeface that has a distinctive character, such as Comic Sans or Old English Blackletter. Suddenly we realize that the idea of an allographic system is also problematic—information has been lost and gained in the translation, even though both fonts are allographic in Goodman's sense. The specific properties of instantiation are not just embodied in the letter code, but in the inscriptional form of the characters. All graphical notation systems turn out to have some aspect of the autographic in them, no matter how minimal, and the loss of information that occurs in remediation is considerable. Whether it matters or not will depend upon the purpose to which the graphical image is being put. I would argue that truly allographic systems do not exist, for the reasons mentioned above. A system may be allographic at a formal level, as a notation system, but never at an inscriptional level where an image is produced as a material trace.

In *The Semiology of Graphics* (1967), Jacques Bertin outlined formal features of graphical systems from a different perspective.[44] Bertin's motivation was directly related to problems in cartography and the need to create a rational approach to the use of graphical features to communicate large amounts of information in a clear, legible way. He identified seven graphic variables—color, shape, tone, texture, orientation, position, and size—so that these could each be put to a distinct and deliberate purpose in map-making. Shape communicated semantic values, such as labels or types of information, more clearly than tone, which was better suited to comparison of intensities of value. His system was created to guide production of static graphics, not dynamic ones, and did not include any of the properties related to animated images, such as change over time, movement, growth, and so on. But the principles of his approach remain useful in design—as do those of many information designers from earlier generations whose manuals and/or examples are part of the inventory on which we draw.

With these thoughts about graphical knowledge in place, a few notes about the way visualization is used in computing might be useful in thinking about how the distinction between modeling interpretation and visualizing knowledge are directly linked to the challenge offered by *graphesis* to *mathesis*. What are the technical conditions and critical implications of modeling interpretation? Where and how are data structures to be created, stored, made flexible and iterative? What kind of information and knowledge are they able to encode? How can they remain sufficiently structured to function within the formal systems on which computation works while also inscribing hermeneutic specificity?

Philosophical issues of imitation, mimesis, and simulation
The development of digital image production intensified, rather than eliminated, philosophical debates about the epistemological stature of images.[45] Questions about the epistemological identity of digital images were posed in terms of what their relation was to coded files—or simply, to code itself. Are digital images equivalent to their digital encoding? Or do all of the various processes of encoding, processing, and display need to be taken into account as forming a composite or aggregate identity? What, to put it bluntly, *is* a digital image? What constitutes it?

In forensic terms, as Matt Kirschenbaum has eloquently demonstrated, every digital trace is unique by virtue of its physical materiality.[46] While on a formal level, as a logical expression, a string of code can be repeated. But the instantiation of that code in a particular location on an actual digital device of any kind creates a specific trace that is unlike any other. The extent to which this uniqueness constitutes the epistemological identity of an image file may depend on whether an ontological question is being posed, or a pragmatic and operational one. Is a digital image, in its instantiation, unique and specific? Or is its potential for instantiation repeatable and/or iterative?

Hand-drawn images make truth claims based on verisimilitude (accuracy) and resemblance. A face drawn in an illustration program may resemble its source, but it does not bear any direct connection to that source except through comparison of visible features. Photographical images, however, base their claims on their production as direct traces of light. A scanner also creates a direct trace registering light in code. In either case, the file stores the information in a formal notation system, universal and context-independent, even if each inscription is distinguished from every other. But is the encoded file related to a display through an indexical chain? What ontological, or basic relation of identity, do the file and image have?

This contrast was noted in 1996 in an article by Hubertus von Amelunxen on digital photography. Amelunxen described two types of mimesis in digital images.[47] He distinguished direct indexical traces (likeness/eikon) in traditional photography from the manufacture of images that follow codes of verisimilitude (simulacra). In a digital environment, both types of image exist. Photographically naturalistic works need not have a referent to pass as images of the "real." But no matter what its source—camera eye, mouse, or software—at the point at which data is inscribed, it has a direct, specific authenticity as a forensic trace. The encoded trace has an ontological identity as a file.

In data display, an indexical link exists between the direct trace of light and its encoded record and also as a connection between code and display. None of these are equivalents. They are various iterations and remediated translations. This suggests that a claim can be made for the truth value of digital images as expressions of encoded data. The data and image are not one and the same—data can be given any number of graphical expressions, played as music, computed, rendered in color or black and white, and so forth. But the digital image is (popularly and fundamentally) conceived as a truth of another kind that is premised on a deep conviction about the relations of reason and truth, a rational link between mathematics and form, in which the identity of a mathematical formula is supposed to exist irrefutably, absolutely, as an indispensable truth. This idea of truth in digital images seemed unsupportable then, and now. The truth invoked in a visual expression is not that of the data. In this positivist premise, the foundation of a digital ontology is linked to a belief that mathematical code storage is equal to itself, a truth that is based on identity irrespective of material embodiment. Once produced, data have a cultural authority that masquerades as an inherent (ontological) authority, pretending to an absolute self-identicality. When images such as visualizations are taken as equivalents to or for data, confusion arises.

Visualizations are not automatically generated from data. A single data set can be expressed in any number of graphical formats. These images stand in relation to the data as "Copy" does to "Idea" in a Platonic scheme. We might even argue that the visualization comes closer to Plato's more debased category of the "Phantasm"—which is a copy of a copy—if we consider that the data are an abstraction from an original phenomenon.[48] The fact that data, no matter how cleanly and formally they are encoded, have to be processed to generate a visualization, makes it impossible to assert any direct identification of code and image. These are (ontologically) distinct entities, linked in a relationship to each other that is indexical by virtue of its continuity and possibly iconic by virtue of the formal, logical structures encoded in digital files. Data and expression have a clear indexical link; they are not equal to each other.

In addition, in answer to the claims of "pure difference," it should be noted that code has no ideality, no independent (or transcendental ontological) existence. Code is firmly, resolutely, and substantively material. The visual forms to which data give rise, its manifestation into substance, is what allows it to be available to perception and cognition. In a weak analogy to snowflakes, or some new age Heraclitan observation, it is fair to say that no two pixels are alike and that instantiation always bears in its material embodiment the specificity that makes for difference from the code—and from the data. In the visual practice of an information design, the assumption is that the information precedes the representation, that the information is other than the image and can be revealed by it, served by an accurate visual presentation. But form is *constitutive* of information, it is not merely its transparent presentation. The very acts of production and inscription, the scribing of lines of difference that create the specificity of an image, demonstrate the making of the form as an act of differentiation from *mathesis* (code). Furthermore, code, however conceived, cannot be conceived as "pure" if purity suggests some independence from a material substrate or instantiation into material. Code is also, always, emphatically material, not pure.

However we conceive of the material inscription of digital information, in its logical and formal structure, it has an unambiguous character at the level of binary code. Rather than think of this as "pure difference," the encoding should be thought of as materially dependent inscriptional specificity. This is not meant to suggest that code is self-identical, or that it has an "ideal" form to it as an expression of "truth" even at the level of inscription. Once expressed in graphical form, the highly discrete and particularized aspect of inscription guarantees its distinction in every instance. Another important point is that visualization does not depend upon the preexistence of code. From an early period of computer graphics, in Ivan Sutherland's design of Sketchpad, we witness the demonstration of the direct production of code *from* image, of data *from* expression. This is significant, since the process of transferring information from a hand gesture to a screen to a file is always a mediation, with the attendant loss and modification of the information in the original inscribing act. No amount of computational power can fully record every nuance of difference in the pixels of the screen—for no two pixels are the same, and no pixel is the same as itself over time. In other words, digital displays are inscriptional by virtue of their material instantiation.

This discussion of the connection between digital images and their material instantiation leads to an examination of the specific properties of graphical inscription. The distinction between the formal system of code and the specificity of each instance of instantiation further raises questions of what constitutes the "information" to be encoded in either a direct production of a graphical production of knowledge or a remediation of a preexisting artifact and/or expression of a data set. If "form" is conceived in mathematical terms, essentially formal and logical, it can be understood in terms of a unity of essence and representation where conceptual value and expression are the same. But if "form" is conceived in terms of *graphesis*, then it resists this unity in part through the specificity imparted by material embodiment. This materiality cannot be fully absorbed into (or made one with) form as "pure code."

The gap between the formal language of code and the particular, unique, specificity of any inscription marks the difference between the claims of *mathesis* (as universal and absolute) and those of *graphesis* (as located and infinitely varied). This distinction can be expanded into an argument for inscription as a fundamental aspect of *graphesis*. This situates any graphical act, whether generated from data as display or created directly on screen or in other media and materials, within cultural and social systems where the condition of materiality permits and/or requires critical considerations of the ways material form participates in and helps replicate cultural mythologies. In the case of digital images, this is a mythology in which code passes for truth, as if the easy and complete interchangeability of image into code and back into image is driven by a myth of the techno-superiority of mathematical premises. The idea of *graphesis* makes an argument against the concept of code as truth, by insisting on the distinction between *the form of information and information of form-in-material*. As a materially inscribed trace in a complex system of interrelated components, processes, and operations, code cannot escape its relation to *graphesis*. Code is always an inscription (an act of primary making) and inscriptional (characterized by the attributes of specificity and particularity forensically inherent in any trace).

This theoretical framework has been essential for establishing the fundamental identity (or ontology) of inscriptional forms within digital environments, and for making an argument for codes as inscriptional traces. But it does not provide any substantive description of the specific capacities of graphical forms to produce, instantiate, and communicate knowledge at a formal level. So let's consider some of the ways graphical notation has been analyzed and its properties described.

Visualization methods used in computing

The previous discussion approached the status of digital images from a philosophical perspective, but the pragmatics of technical process must also be taken into account in any discussion of visualization. This aspect of the way knowledge and interpretation are produced occurs at the level of the workflow that creates what are commonly known as information visualizations. Computational techniques linking data and visual expressions constitute a specialized field. Its methods are created with specific uses in mind that shape the data and their display—and the relation between the two.

Standard visualization methods are performed from data through five distinct methods of algorithmic interpretation:

1. *Description*: A descriptive presentation of already known data in a visual form allows patterns of large quantities of information to be seen and understood. The quantitative processing of massive amounts of information supports an exponential leap of qualitative judgment. The benefits of analysis and visualization are most evident at a large scale. (A problem like showing the comparative length of Shakespeare's plays and the number of major and minor characters in each can be best understood through the compact presentation of information in a chart.) This is the standard graphics-expressed-as-metrics approach to visualization. It is pervasive, and so common that it is barely questioned. But this approach actually makes declarative statements from probabilistic models, that is, it is using *capta* (the products of a process of parameterization and modeling) as if they were *data* (preexisting entities). The longer lifecycle of data production is not visible, and the values expressed seem unproblematic.

2. *Analysis*: Analytic visualization allows for querying of data to see if it meets certain conditions, displaying those sets of information that conform to the limits (or parameters) of the query. Data, already present, can be analyzed and visualized. (For example, we could ask how many of Shakespeare's works were written during any particular time period and how often they were performed.) This approach also uses graphic display as if it were a clear, reliable, unproblematic statement of data.

3. *Modeling*: In a radically different approach, modeling in visual form allows a hypothesis to be tested through visual means. Imagine, for instance, creating a model of a molecule, studying the resulting characteristics, and then putting another chemical substance into that configuration to see what properties it begins to exhibit. The same procedure could be performed on an aesthetic artifact by imitating formal features of meter, rhyme, narrative, semantics, or word frequency abstracted into a structure that is visualized and used to generate another work, etc. In this approach a higher-level formal structure embodied in a graphical expression is used as a template.

4. *Procedural approaches*: By contrast to the above modes, procedural visualization begins with an algorithm that generates form through a series of evolutions. (Process driven works of literature use such methods to create works from, for instance, vocabulary lists and rules for their

combination. Permutational and combinatoric works also fall into this category.) The graphic is generated directly from the process, or set of instructions, and used to read the outcome or result while concealing the procedures in the algorithm.

5. *Emergence*: Finally, ways of making emergent models rely on developments and mutations that allow an algorithm to self-modify according to criteria of selection that simulate evolutionary models. (The rules of the procedure modify themselves so that not only the forms, but the terms on which they are created, continue to change.) These emergent models, such as those in agent-based and nonlinear systems, make use of the graphic display as a window into the unfolding process.

These operations are used for data manipulation and can be applied to texts as generative and analytic processes. We could argue that each of these is, in fact, interpretative on account of the model underlying its operation, but only the third approach, modeling, allows a user to generate a graphical intervention for the purposes of creating an argument. To address that problem, we need to push further. In addition, we need to consider the critical understanding of these approaches. As most of these visualizations come about as the result of procedural, mechanistic approaches, we have to consider the way they participate in what we might term *the graphical imaginary*, the realm of ideological and cultural knowledge produced through these symbolic systems. Images produce values and/as knowledge, they combine concepts with ideologies, and never more so than in the computational environments of current culture, with the heavy dependence on graphical formats in mediated communication cycles of everyday life.

Summary

A basic (epistemological) distinction can be made between visual notation and inscription. On this basis, the specificity of inscription is understood to challenge generality. Unique instantiation pushes against formal systems that claim to operate in a context-independent manner. Understanding this is essential if visual methods are going to be used to serve interpretation and user-dependent knowledge production. The very specificity of inscription gives graphical forms their identity, fundamentally (ontologically) and materially (pragmatically), and makes them well-suited to assist in the production of models of interpretation through graphical means, or as expressions of discursive, expressive forms that cannot be subsumed in formal, logical systems. This distinction brings us back to another basic issue—whether graphical forms are images *of* something, and thus stand in a secondary or surrogate relation to a preexisting form of knowledge, or whether they *create* knowledge in a primary sense. Because, in a digital environment, this question is often posed within assumptions about the character of code, insisting on the inscriptional trace is crucial.

In the rhetoric of cyber-speak, where data has somehow mistakenly come to carry an aura of immateriality as a feature of its fundamental identity, that identity has come to be conceived in a relation of identicality—of information to itself. Concepts of identicality—of something that presents a closed identity, as if it were in a stable and constant condition of being—are problematic from the perspective of the politics of representation. They lean toward totalization as a model of thought that leaves little or no room for the critical action and agency that are essential to any political basis for agency.

Central to the argument here is my conviction that an ideological agenda underpins the concept of self-identicality, as well as the cultural authority of formal, notational, systems. The specificity of graphical modes of expression presents an argument against the very possibility of such identicality and the truth claims it embodies. *Graphesis* can be positioned to challenge *mathesis* through a discussion of the ways the ontological identity of data visualization is generally conceived. In other words, in physical reality, nothing is ever the same as anything else—and may not even stay identical to itself moment to moment. Part of this assertion depends on a careful characterization of the way graphical expressions are instantiated in material.

Within the digital humanities, what is at stake in this discussion is very simply the question of *what kind of knowledge* we are engaged with in our work. The crucial epistemological issue in advocating visual, graphic expressions for interpretative work is based on the distinction between inscriptional specificity and notational generalizability. Specificity suggests a unique, even if non-self-identical, expression of an interpretative act. By definition, this is interpretative because it does not claim truth status as universal, repeatable, or independent of the circumstances of its production. Inscriptional expressions—whether analogue or digital—retain the trace of their conditions of production as part of their forensic identity. These conditions are embodied in the marks of their production and are thus able to be read as indexical traces as well as contributing to the semantics and rhetorical force of argument and evidence. By contrast, the generalizable character of notational systems exempts them from these considerations. Notation functions as a user-independent means of encoding information, knowledge, experience, and so forth in a system that can be (ostensibly) decoded without regard for the specific qualities, or *qualia*, that are inscribed. In other words, an alphanumeric string can be used, copied, and remediated and still be functional; it does not have to be made or drawn in the same manner each time to remain useful.

If the humanities are attached to the notion of qualitative, subjective, located, and particularized expressions of knowledge as experience and evidence, then the recognition of the unique role of visual epistemology becomes clear—for the visual argues in its very expression and instantiation for the located particularity of human knowledge. In direct contrast (not opposition, but distinction) to user-independent knowledge, visual inscriptions argue for the specificity of user-dependent conditions of knowledge production as interpretative—where interpretation signals the subjective, located, inflected, and particular character of knowledge located within a subjective experience.

This argument is not meant to suggest that visual systems cannot be consensual, or that common experience cannot be communicated, or that standards and standardization cannot be expressed and achieved using visual means. All of these are well within the realm of visual epistemology. But they are not what distinguishes visual forms of knowledge from others. In a similar way, when dealing with sound recordings of speech, we note that the inflection of semantic value depends upon vocal expression as an inscriptional trace. This is inseparable from the conditions of production and reception. Gesture, voice, tone, style, and organization of print and layout, design and formal properties, all participate in the production of semantic value. They, too, are inscriptional in their character, not notational. They are specific, not general, particular not generic, and rooted in a user-dependent condition of production and reception, not independent of it. This locatedness requires attention and critical engagement in reading inscriptional materials.

Visual epistemology is distinct in the properties it brings into knowledge production—those of resemblance, iconicity, spatial arrangement, and so forth. But it also has properties that can be compared directly or by analogy to those of hearing, such as tone, color, pattern, frequency, and order. The ability of the eye to make and register minute distinctions is remarkable, and much untapped potential to engage this capacity in graphical expressions can be pressed into the service of interpretative work. These can be far more nuanced and subtle than those in current use in visualization conventions.

No technical obstacle exists to recording, structuring, and making this kind of inscriptional work into data, and the challenge is merely one of sorting out what features of the visual are useful in creating arguments and interventions within the realm of digital projects and platforms. What matters is not what kind of epistemological contribution is made by visual means, but how visuality models knowledge, or epistemology, according to particular parameters of inscriptional specificity and notational generality.

Next

Before turning our attention to critical interface design, nonrepresentational strategies, and a suite of projects in which these theoretical principles are embodied, we will pause to consider the fundamentals of critical hermeneutics recast in a performative and probabilistic mode. To fully explore the idea of *graphesis* as a means of producing interpretation, new knowledge from a subjective point of view, we need to extend the concept of knowledge, or epistemology, to include a probabilistic dimension. We have to go beyond thinking of knowledge in terms of mechanistic and static relations in which things known and things shown are assumed to be independent entities operating in an objective universe of phenomena existing in advance of their apperception. Visual epistemology is based on a theory of knowledge as a codependent production of subject and object in what Karen Barad has described as an *intra-action*.[49] This assumes two entities exist in relation to each other but are, crucially, also part of the same system. Their codependence cannot be disentangled. The process of interpretation that is at work in probabilistic hermeneutics draws on concepts of constructivist engagement that erodes subject-object distinctions. Now it is time to discuss these concepts in more detail and see how they relate to or are served by graphical expressions in a digital environment.

2   Interpretation as Probabilistic: Showing How a Text Is Made by Reading

As we have seen, knowledge and interpretation can be produced and represented in visual forms and formats. Graphical forms can structure arguments and express rhetorical positions. But can they be used to expose the workings of interpretation—show its models, assumptions, and operations? If so, through what means and with what benefits for humanities scholarship in its ethical as well as intellectual dimensions?

Interpretation: A probabilistic approach

If we are to model interpretation, we must understand what model *of* interpretation we are assuming. In the twentieth century, interpretation, or hermeneutics, went through what scholar Gerald Bruns has termed "a Copernican revolution"—by which he means the transformation of a user-independent model of textual meaning into one centered in individual understanding.[1] Discussing Martin Heidegger and his profound influence, Bruns describes the shift from positing the site of meaning *within a text* to identifying it with the *act of reading* as a process of negotiation.[2] For Bruns, the core concept within this tradition is that of *understanding*, which is to be conceived positionally *and* intellectually. *Understanding* emphasizes process, the state of being in a relation to a work (text, image, experience), rather than arriving at closure of meaning or sense.

The concept described by Bruns became the basis of critical theory in twentieth-century literary studies. A series of critical developments—reader-response, deconstruction, psychoanalytic theories of the subject, the play of work-text-word-trace, and other constructions of the dynamic production of text in the complex and situated act of reading—became the paradigms of interpretative, hermeneutic practice. The notion of an essential meaning that could be divined from a text, to be extracted or exhumed, revealed or taken, was replaced with the idea that the reading subject occupied a position in relation to a work, and that the position was informed by the subject's own lifeworld, knowledge, and conditions of identity across every possible and conceivable factor. Our positionedness and our subject identity were a complex of dynamic systems that factored into the relationship of reader to text. Identity politics of authorship and readership became essential features of critical work. The larger concept of decolonization of knowledge is the view that knowledge needs to be subject to critical assessment that takes into account its relation to the power systems in which it was produced. Decolonization thus embodies political impulses that have been fueled by historical insight and ethical reflection on traditions of cultural practice and their legacy, particularly in the west.

Decolonization draws on political critiques of unacknowledged bias as well as legacy assumptions structured by power relations of colonizer and colonized. These legacies have long histories and become almost invisible within the dominant (or hegemonic) practices of discourse formation. Subject positions imprinted through education and acculturation become blind to their own circumstances. To engage with understanding in the twenty-first century requires awareness of these formative conditions as part of our individual subject positionality. These tectonic shifts are marked in literary discourse and analysis as well as professional practices in information management and organization, but digital tools to inscribe the positionality of historically situated work (and subjects) have been slow to develop. Decolonization of knowledge is a strategic move and requires breaking the singularity of point of view often enacted by hegemonic discourse. Fracturing, faceting, and multiplying positions of cultural authority are crucial to this activity and builds on critical principles of hermeneutics, extending them into a dialogue with the political aspects of interpretation and/as knowledge.

The now-common conception of critical interpretation, or hermeneutics, acknowledges that a text does not transfer meaning in any mechanistic way, but it still stops short of the recognition that all acts of reading, productive and constitutive, are probabilistic—by which is meant that they are nondeterministic and selective. A text provokes a reading, allows it to come into being. Every reading is specific to its reading moment, it is an event generated between a reader and a text in a set of conditions never to be repeated.

The range of potential readings creates a field of possibilities, and every act of reading is an intervention in that field, a moment at which the generative potential of the text is intervened, flattened into a specific reading. A normative distribution applies here, with the bulk of readings of a given text clustered together and outliers stretching the field. But that hardly matters, because even if the readings aligned, were so close as to appear almost the same, minute examination would reveal the inscriptional specificity of each one. No reading is ever the same as another, and no text is ever equivalent to the interpretation it provokes. Inscriptional specificity, as described in the previous section, applies to experience as well as representation.

We arrive in the early twenty-first century, therefore, with a clear recognition that knowledge, as well as processes of knowing through interpretative work, is situated, partial, historical, and cultural in its formation. Individual readers engage in textual interpretations in accord with their training and background as surely as the texts are also the expression of specific circumstances of production. This recognition does not, by any means, reduce hermeneutics to demographics, nor texts to symptomatic expressions; quite the contrary, it simply insists that the start point for the production of a work through an interpretative (hermeneutic) encounter is always a complex act of codependence.

As my former collaborator Jerome McGann put it in 2001: "a literary work codes a set of instructions for how it should be read. Unlike machine and program codes, however, these codes decipher to an indeterminate number of precise outcomes. They represent exactly what Jarry called 'a science of exceptions.'"[3] McGann drew on the work of late nineteenth-century poet Alfred Jarry, who defended statistical anomalies as the foundation of the field he called 'pataphysics.[4] In the context of SpecLab, the theoretical digital research group McGann and I conducted in the 2000s, I coined the additional term "patacriticism" to gesture toward the use of Jarry's concepts within our research.[5] The notion was that statistical norms and procedures of interpretation had to be balanced with other probabilistic considerations, among them the recognition that outliers were to be taken as seriously as acts of "deformance" (deliberate distortion or transformation of a work), along with other interpretative techniques.

Our approach at SpecLab focused on the probabilistic character of reading within the design of Ivanhoe, a game of interpretation designed explicitly within the digital humanities, on which more will be said below. The challenge we grappled with still remains: to model the basic principles of a probabilistic approach and give visual expression to its constructedness, partial understanding, and the situated condition of interpretation. In this section, several concrete suggestions will be discussed for using graphical means to (1) break the apparent singularity of statements, (2) introduce comparative methods that expose assumptions, (3) demonstrate the madeness of information and graphical expressions, and (4) embody interpretative principles in the design of platforms. The goals of these interventions are value-driven (ideological) and ethical as well as intellectual.

Digital humanities and probability

Digital tools penetrated the humanities with increasing impact from the 1970s and '80s onward.[6] They were part of the world of information professionals much earlier in data processing and records management, and some of the early repository building and text markup tools crossed between the management and research domains. However, the tasks taken up in digitization not only did *not* challenge computational techniques, but submitted to them, albeit often with considerable reflection and discussion, much of which ended with a shrug and then submission to the demands of computational processes at the price of theoretical concerns. Text analysis made use of searching, matching, counting, sorting—all methods that rely on using nonambiguous instances of discrete and identifiable strings of information coded in digital form. This kind of automation depended on principles that seemed to be nonhumanistic. The creation of a concordance, as in the oft-cited case of Father Roberto Busa's pioneering work in the 1950s on Thomas Aquinas, does not, on the surface, appear to require interpretation. The identification and counting of particular words seems simple. But a word does not have a single value or meaning. Even identifying a term requires selection—should it be just the word, its variants, the word with one or two words collocated around it, and so on. That such work is necessarily interpretative, that it actually depends upon a *model* of the work being done, was a point carefully left aside. The results of the automation and insights it provoked were a tool for analysis of Aquinas's vocabulary, but still had to be interpreted. But the work and its methods did not produce a challenge to computation or call for extending its capacities to engage with ambiguities and contradictions.

By the late 1980s and early 1990s, humanists were engaged in critical editing and repository building using digital tools and then networked environments. Metadata (information about data) and markup languages (formal tag sets inserted into a text for purposes of interpretation) are both highly structured forms with professional communities and standards. Working with them forced a dialogue between traditional humanities and the formalized information structures of content types and data. The mantra of the period—that the engagement with digital tools forced humanists to make explicit many of the assumptions in their work that had long remained implicit—was complemented by another observation: that humanists started the design of their digital projects as relativists (filled with the lessons of deconstruction, critical theory, poststructuralism) and then gave in to the pragmatic exigencies of positivism just to make things operate. Introducing ambiguity, for instance, at the level of markup languages, metadata, or other data structures, simply did not work. Or so we were told and came to believe.

Formal languages have the capacity to encode complex and ambiguous multiplicities of meaning. The problem of encoding ambiguity, for instance, is not technical. But at the level where project design was being governed by protocols that required explicit and unambiguous terms of operation, ambiguity could not be permitted. No technical obstacle prevents overlapping hierarchies in markup provided each interpretative frame is stored on an independent layer, for instance—a simple solution to a complex problem.[7] But overlapping hierarchies, a much-discussed problem in the 1990s use of markup code, could not be addressed easily within the then-existing tools. More sophisticated models and tools have been developed, but these have not resulted in the creation of new data structures, just means of storing and processing markup. The challenge of modeling tools from humanities principles has been engaged in a few instances (such as NodeGoat and CATMA, both tools meant to support more intuitive and flexible interpretation in digital environments). But for the most part, researchers in the digital humanities have proceeded to develop standard platforms (or adopt them) without pushing humanistic concerns into the computational operations. Even basic techniques for being able to put multiple values into a field or create a metric based on affective values have not been designed into digital humanities systems. Proposals for this kind of modification are systematically integrated into the projects described in the final chapter of this book.

In asking how digital techniques would change, therefore, to embody these interpretative dimensions, I assert that humanities approaches would proceed from a number of specific principles. The first is that interpretation is performative, not mechanistic—in other words, as noted above, a text is not self-identical, each reading produces a text; discourses construct their objects; texts (in the broad sense of linguistic, visual, acoustic, filmic, dramatic, even architectural or site-specific works) are not static objects but encoded provocations for reading. Finding ways of showing these principles informed our discussions at SpecLab, in particular in trying to design the Ivanhoe platform. But the fuller project of showing interpretation, modeling it, making a composition space in which ambiguity and contradiction could coexist, where the non-self-identicality of objects could be made evident within their codependent relation to the social fields of production from which they spring (a relation premised on the constructedness of their identity, rather than the relation of an "object" to a "context") remains unrealized. The caveat that an instantiation of an interpretative act might reify it in ways that create the illusion of fixity does not obviate the need to push for these experiments with humanistic precepts. Especially since the earlier phases of critical editing, repository building, and markup have been augmented by analytic processes of all kinds, with their dependence on visualizations. The tools of visualization, in particular, depend on statistical and quantitative methods from other disciplines. Most significantly, they depend upon explicit parameterization. But almost no humanistic document or discourse lends itself to such parameterization. For instance, how do we date documents—by year of conception, execution, publication, reception? Or how do we characterize language as gendered? What are the moving targets of meaning and semantic value as the use of a word or phrase shifts across contexts and populations? How do you identify a sentiment consistently? The basic impossibility of creating metrics appropriate to humanistic work motivated my argument about the need to distinguish the constructed-ness of *capta* as an alternative to assumptions about the given-ness of *data*.[8]

The argument against quantitative reductiveness is not a dismissal of statistical methods, quite the contrary. One of the problems of digital humanities research has been the use of counting methods (how many instances of X appear in a corpus) rather than statistical ones (given that X appears, how likely is it that it will appear again?) in shaping projects. Statisticians are concerned with probabilities, not certainties. They do not merely count things;

they model conditions and possible outcomes. In an interpretative model, what disposes a reader to assign value to a word or image? Data mining in the humanities has largely depended on automated calculations and simple counting. Statistical modeling has factored less in the analytic toolkit of humanists than in social sciences. Stylometrics, attribution studies, natural language processing, and other higher level analyses have long made use of statistical techniques.[9] But even when these are used, the design of graphic conventions for showing ambiguous, contradictory, or partial knowledge is in early stages, where it exists at all. Work in archaeological reconstruction, where various models have to be extrapolated from partial and fragmentary remains, has created spectral palimpsests to portray degrees of certainty. But even highly speculative economic, climate, or population models have not pushed the development of graphical methods that can fully serve to present their basic probabilistic premises. Humanists have not stepped in to fill the breach, even though the interpretative methods that lead their work require it. We need to develop an inventory of techniques for indicating, for instance, the distinction between what is known and what is projected, pronouncements linked to evidence and speculative rhetoric. How can we show how we think—and how the objects we think about are constructed in the process?

Because humanistic theory provides ways of thinking differently, along lines of interpretative knowing—partial, situated, enunciative (speaking and spoken positions), subjective, and performative—our challenge is to take up these theoretical principles and engage them in humanistic methods of production, *and the production of these methods*. We need to create ways of doing our work that allow us to display its models—not merely modeling a text as if it were a given, or as if it were self-evident, but modeling our reading and interpretation of it.

Experimental approaches to the use of visualization take two varied approaches—data display and modeling. Experiments in data display focus on improving discovery tools, using ways to filter data in various ways that can be made legible by the visualization. The distinction between display and modeling also reinforces the differences between concepts of data and capta.[10] In these circumstances, data are considered objective "information," but in fact, data is information that is captured because it fits the model of what is being measured or parameterized. In other words, all data is actually capta. The data does not exist independently, but is captured as the result of the parameters of the search. The distinction between data display and modeling interpretation is more difficult to maintain in practice than to define in theory. Because data are based on models, we tend to see what we look for in accord with interpretative agendas. Our models continually change, however, and are often iterative in a hermeneutic practice. For instance, the plausibility of dinosaur imagery in popular imagination has a direct relation to the use of increasingly photographic properties of realism. These depictions make the dinosaurs believable because they conform to visual codes that seem linked to the observation of real objects—but then our idea of what a dinosaur was becomes imprinted by those visual images so we look for evidence to match these. Scientific observation follows constructs and concepts. The new vogue for feathered dinosaurs is the effect of a new model based on interpretation of new and older evidence combined. This is usual. Similarly, terms like "terrorism" and "hacktivism" emerge and become defined through use, then seem to refer to an actual category in the world, not merely a term in discourse. Knowledge formation is always a shifting realm of concepts, themselves part of a larger stochastic field of cultural processes. Disciplinary and cultural shifts in knowledge are marked in and effected by such changes. The notion that knowledge is a process of understanding, not an apprehension of things in the world, arises from a constructivist approach that engages probabilistic models. Theories of subjectivity form a crucial component in our current conception of interpretative knowledge as understanding. Without it, knowledge production and representation in the humanities—in any field—will remain locked into mechanistic models of thought in which an image/text is "out there" and an eye brings it "into" the mind. Approaches to knowledge that draw on complexity theory extend models of emergent behavior in agent-based projections and systems that model nonlinear processes. These systems could be used to model graphical expressions of interpretative capta as something that arises from a process rather than existing in advance.

Subjectivity is a structuring principle, not just an inflection. Used in the theoretical sense developed in linguistics and psychoanalysis, subjectivity refers to the place of the observer within phenomena. Subjectivity is conceived as a dynamic, codependent system of relations between the observer and the observed. But though this works at the level of individuals, subjectivity is always situated in social space. To cite McGann again, "Stanley Fish's concept of an interpretative community is a device for measuring the probability function of different interpretive acts. How those probabilities emerge—how certain acts of interpretation gain authority—is a problem that will have to be addressed by studying the normative dimensions of textual fields."[11]

Early in the twentieth century, physicists recognized that empirically based concepts of natural laws grounded in mechanistic, Newtonian models were not sufficient to explain many physical phenomena.[12] Quantum physics and principles of uncertainty characterized a radical change in the ways the observer and observed phenomena are understood, collapsing the two into a dynamic system of codependencies. Rather than imagine discrete phenomena available for independent observation, or the subject-object relationship as a dialogue between two independent entities, the quantum theorist suggests that phenomena arise when an observer intervenes in a field of potentialities. Before that intervention, phenomena may exist in an indeterminate condition, between two states. The intervention forces a resolution (this is the famous experiment known as Schrödinger's cat, named for the physicist who invented it). Probabilistic methods based on the same principle belong squarely within the realm of humanistic methods, but they have rarely been invoked.

Theories of radical constructivism (a theory of knowledge that posits a codependence between phenomena and knowledge) and cognitive studies provide additional disciplinary frameworks. These stress dynamic, relational, systems-based, emergent concepts of knowledge that are as far from naïve empiricism or behavioral psychology as Heisenberg's uncertainty is from Cartesian rationalism and Newtonian physics.[13] McGann sketched this theoretical intersection in his 2001 "Texts in N-Dimensions" composed at the time we were designing Ivanhoe. His approach to texts combined theories of knowledge and cognition to make use of quantum theory as part of interpretative practice and the study of texts. This probabilistic approach to interpretation extended reader-response theory into a dialogue with indeterminacy as conceived in early twentieth-century theoretical physics. Though our work in the humanities was being done a century later, it still provoked skepticism.

However, with this framework in mind, we can move on to describe ways to use graphical forms for the self-conscious creation of knowledge. In other words, we can outline ways to use visualization to create, show, and compare probabilistic hermeneutic approaches. Our first area of focus will be on the principle of constructedness that emerges from the distinction between data and capta and the ways visualization methods can be used comparatively to expose the interpretative dimensions of quantification (production of numerical values) and parameterization (specification of values to be captured). This discussion lays the foundation for attending to the problems of argument structures in a discourse field, the partial and situated character of interpretation, and basic operations of hermeneutic work.

Constructedness: Visuality and probability

Each form of expression, and each medium, has its own properties, and these require specific interpretative approaches. Architecture cannot be analyzed in the same way as music, for instance, no matter how many comparisons can be made, and likewise, a poem cannot be interpreted through attention to the same properties one looks at in a dance or an image. Visual works (and film, performance, and other materials) lend themselves to some of the same critical practices as text, but only in the sense that all aesthetic objects are engaged in a similarly probabilistic manner. The specificity of visuality comes into play in interpreting visual work because of the graphic properties of composition, iconography, style—and the instability of visual signs. The training to read the codes of visual images is distinct from the training to read texts, and the histories of art history, visual communication, and web design are full of methods appropriate to understanding visual symbols and structure. Similarly, specialized training is needed for reading aerial photographs in military surveillance, sonograms and MRIs in medical research, or any other graphical expression used within professional domains. As per the earlier discussion, no "innocent eye" exists, and no images can be apprehended directly without some amount of expertise. Human beings in varied cultural circumstances may not even recognize an image as a visual representation—the basic category has to be learned before it can be understood. The interpretative engagement with visual forms is therefore as highly mediated a reading practice as that of texts, formulae, musical scores or any other encoded artifact of human expression. Critical intervention in a visual discourse field is at least as complex as it is within the nuanced subtleties of a textual one.

But in this context, our focus is on how visual formats can be used to *show* models of interpretation in three areas of activity: data/capta production, situated authorship, and argument structures. Can we expose *models* of interpretation in these activities and give them graphical form as a means of apprehending specific features and properties?

To begin, we need to reiterate the distinction between *data* and *capta*. *Capta* is created from phenomena, abstracted through parameterization into what becomes reified as *data*, and then engages larger structures of interpretative arguments constructed on and through a rich discourse field. One of the primary ethical issues in such practices is to dispel the assumed neutrality of data production. Can visual means be used in part to locate the situated authorship of data, the position from which it is created, spoken, and then used as if it were value-free? A crucial task is to expose the myth of data as "given" and observer-independent. The fundamental interpretative act of data production is often concealed, in part because no conventions exist to display its processes, and in part because the intellectual task is often discounted. A second major issue is to demonstrate the situatedness of authorship and readership. This is fundamental to humanistic research since it locates scholarship and argument within communities of practice and individual circumstances. Finally, how can we call attention to argument structures and the way they make use of linking, organization, and analysis of evidence in the form of documents, artifacts, materials, sites. The extended use of analytic derivatives and surrogates adds new dimensions to historical and cultural values within argument structures, and these should be signaled as well.

To show the process of *capta* creation means creating a series of discrete steps for depicting the way statistical operations shape the production of data from phenomena. What, for instance, is the initial material from which data are produced? How is the sample size represented in relation to that original phenomenon? How can specific decisions about parameterization (specification of what can be counted) be revealed? Two approaches to answering this question present themselves immediately. One is to show the decisions built into the lifecycle of data production, the other is to show the nonsingularity of any presentation of *data* as if it were simply a statement. Both of these approaches can muster graphical means to their cause.

While documentation is essential for showing the interpretative decisions that go into modeling data, a graphical contrast that exposes how the model would change if any single parameter were altered makes the existence of models dramatically clear. For example, let's set the following problem: For each value or parameter used to create a sample set of data, show that a different sample would be produced by altering any single factor in the model. This simple move immediately shows that the data produced by any model is only an expression of that model and that the visualization is of the data model, not the phenomenon from which it was extracted. The outcomes can be compared graphically. The contrast of charted information, even in standard formats, makes a striking intervention into the declarative singularity, the appearance of simply stating "what is," of the graphic statement. Singularity is the quality that a statement has when it appears to be indisputable, a statement that merely seems to declare a self-evident fact or observation. One potent critical tool for deconstructing the singularity and invisible authority of any statement is contrast. Comparison breaks the hold of any appearance of singularity of declarative statements in intellectual work and knowledge systems. As soon as such a statement is expressed graphically and the visualization is relativized, shown to be *a* statement about the "data" and not an uninflected statement of fact, then the recognition of the constructedness of the data, its visualization, and the model on which it is formed are all evident. (For instance, imagine changing the scale on a vertical axis in a graph to exaggerate the presentation of difference in value. Such an act demonstrates that the value is encoded in the graphical presentation, not independent of it. The graph is not just "showing" the value, it is making a graphic argument about value in the way it presents the information.) Instead of the usual "this is the data" statement, a comparison of one outcome with another (e.g., stretched vs. compressed scales) makes the argument that "these are propositions about how data might be abstracted and rendered." In addition, the purpose and motivation of the data model should be revealed in documentation—for whom and in whose interests is the model working? Who paid for the work? Why? These are factors that do not rely on visualization, but are crucial in the reading of the outcome and the visual presentation.

Distinguishing data and capta

To overturn the assumptions that structure conventions acquired from other domains requires that we reexamine the intellectual foundations of digital humanities, putting techniques of graphical display on a foundation that is humanistic at its base. This means first and foremost that we reconceive all data as capta. Differences in the etymological roots of the terms data and capta make the distinction between constructivist and realist approaches clear. Capta is "taken" actively while data is assumed to be a "given," able to be recorded and observed. From this distinction, a world of differences arises. Humanistic inquiry acknowledges the situated, partial, and constitutive character of knowledge production, the recognition that knowledge is constructed, taken, not simply given as a natural representation of preexisting fact. My distinction between data and capta is not a covert suggestion that the humanities and sciences are locked into intellectual opposition, or that only the humanists recognize that intellectual disciplines create the objects of their inquiry. Any self-conscious historian of science or clinical researcher in the natural or social sciences insists the same is true for their work. Statisticians are savvy about their methods and about the way choices made in their processes shape arguments and outcomes. Social scientists may divide between realist and constructivist foundations for their research, but none are naïve when it comes to the rhetorical character of statistics as a tool.

Making a distinction between single and comparative statements using graphical expressions is one way to expose the probabilistic character of interpretation. This shows the nonequivalence of different statements at the level of data understood as capta. Showing decisions about parameterization or the shape of data as capta would not guarantee that one or the other had a greater claim to accuracy. Instead, the move would show that each are acts of probabilistic interpretation given graphical expression. The point of such comparisons is to show the constructedness of data, understood as capta. No statement, taken as a single, self-evident expression of data, or of features of phenomena expressed as data, can provide this reflection on the process of its production. Relativizing a statement through contrast and showing this contrast graphically calls attention to the *modeled* nature of the information and its expression.

Take a simple example. Imagine that the occupancy of classrooms needs to be calculated for purposes of space resource assessment. Now, consider the fundamental decisions about parameterization: when do I take the sample, how frequently, at what point in what cycle of periods? Every day at noon? Every night at eight pm? For how long should the period of counting last? Should the sample be taken once at three pm and once at nine am and only on Mondays? What about Sundays? During the entire calendar year?

Or in July? Mid-December? Anyone doing the least bit of statistical research understands the implications of these decisions. No single answer to these questions suffices, but every answer embodies the particulars according to which the "data" (again, my preferred term is *capta*) are produced. Making a series of contrasting calculations and showing them graphically would perform two important intellectual acts—of exposing the decision process of capta production and of demonstrating the rhetorical force of graphical expressions as propositional argument structures rather than as declarative statements. The charts become "what if" statements rather than "this is" statements, and the user-dependent conditions of knowledge production become inscribed in the process of presentation.

The model of interpretation does not simply show the single chart or graph of data points. A chart may express a model, but it is not the model, which is a schematic template that can be used repeatedly. The model consists of decisions shaping the particular parameters or template from which the data or capta arise. A simple but critically revealing exercise to do with a data set is simply to send it through a series of standard graphical expressions—turning it into a line graph, bar chart, pie chart, and/or tree diagram without altering the data set. This makes the graphical properties of standard conventions legible—in part because they are often nonsensical expressions of the information (percentages of opinions held in a population, for instance, do not belong on a continuous graph, with its implication of change over time, any more than marked increases in growth of a species in an area belongs in a pie chart with its static structure designed to show percentages of a whole). These exercises expose the rhetoric of conventions, but not the deeper issue of the way an interpretation depends upon a model.

Interpretative models linked to analytics can be exposed through the basic contrast of alternatives suggested in the two sample exercises just described. But intellectual models are more amorphous and elusive. Topic modeling, data mining, text analysis, and network production, as well as simple collocation and word frequency, are *modeled,* not simply performed. So is markup, and any kind of critical analysis, historical work, or cultural criticism. Providing a way to see the terms of the model, again by contrast, is an essential move.

The modeling of gender provides a much-studied example. Such work has also been thoroughly critiqued by Laura Mandell. She has systematically analyzed the ways gender modeling in digital humanities bears the imprint of unexamined assumptions and their implications that can be exposed *as a model*.[14] What are the specific features of language that are being identified as gendered; how do they vary across texts, conditions, periods, and authors; and what terms of inclusion and exclusion, selection and variation, flexibility and mutability, are being used to create statistical analyses of a text or corpus? Asking these questions and then varying the models to demonstrate precisely that they *are* models, makes an argument that no single statement or presentation can. Breaking the claim to authority embodied in the single statement is a crucial means of demonstrating the constructedness of interpretative work. But also, this points to the situatedness of intellectual work within the varying demographics (population profiles structured according to age, income, race, or other features) and communities of practice, whose interests are also at work in shaping data models and presenting their outcomes.

Graphical formats can work effectively to show and contrast models for the same reasons that they work to expose patterns in data—they are legible and succinct and provide a gestalt overview of large-scale or complex patterns. Only documentation can provide the fuller critical material about communities of research and practice but multiple, layered, faceted visualizations are a tool of critical hermeneutics. They show that every presentation is the outcome of a probabilistic inquiry, a "what if" proposition, not a "what is" statement. This is a crucial intervention in the singularity of authoritative statements of "mere" fact.

Discourse fields

With this discussion in place, we have a positive answer to the question of whether visualization methods in digital humanities can be used to show the critical, probabilistic, and multidimensional features of interpretation. The relation between interpretation and any field of evidence is always partial and incomplete, *a reading through* rather than *a reproduction of* the materials from which an argument is drawn. We need conventions for showing the features of observer-dependent knowledge production that are not part of the history of techniques used in empirical observation, with its assumptions about certainty, disambiguation, and repeatability.

One of the early tenets of belief in digital humanities, associated with the tasks of textual markup, was that the engagement with computational techniques required that humanists make explicit many aspects of their work that had always been implicit—such as the categories on which distinctions and judgments were made. Some of these were relatively straightforward: Does a fragment of text describe an action or an intention? A character's mood or a metaphor of natural forces? A historical event or a personal one? These kinds of decisions govern much of the interpretative activity of textual analysis and markup. When they have to be aligned with hard categories and decisions that assign a value to a word, phrase, or text, then the complexity of embedded and associative properties of language becomes evident. Resolving these dilemmas by forcing explicitness was an early solution. Creating means of rendering the ambiguities clearly and legibly would be a step forward in aligning the digital tools with humanistic methods. The ways the cultural authority of certainty asserts itself, or conceals itself, through digital practices is highly charged. Finding alternatives that counter the certainty of computation with the generative dialogue of interpretation is crucial.

The legacy of critical hermeneutics shifts in the encounter with digital methods. Humanistic approaches force us rethink the compromises these methods brought about through "datafication" of the cultural materials that are central to the humanities and the cultural record. The question arises as to whether aesthetic works require consideration distinct from that of other kinds of works (administrative, fiscal, medical, commercial, etc.) or whether the encounter with these works demonstrates principles useful to all cultural objects. Does a plumber's manual, a medical handbook, an automobile, a subway system switching diagram, a work of technological engineering require the same critical approach as a poetic text? The short answer is simple: not always. (If I am looking through a plumbing catalogue to find a replacement part for a broken toilet, I may not want to engage in critical, interpretative, and probabilistic processes. I might just want to find the right part.) Still, the relevance of critical approaches to all aspects of human expression is one lesson bequeathed to us by anthropology and cultural studies. Of course, differentiating among kinds of texts and their potentiality (and the reader's purpose at any moment) is an important aspect of critical engagement.

But our concern is with the needs of the humanities and the validity of humanistic methods within the growing realm of digital work (work remediated through digitized forms, formats, files, processes, and expressions, usually within networked environments). In this arena, the struggle is not about what objects should come under examination or consideration, but how the critical approach should be formulated to incorporate hermeneutic traditions. Stealth positivism is rampant in digital work, and the unexamined consequences of its use are many. Chief among these, and perhaps the core issue of ceding intellectual ground from the humanities to the methodological instrumentality of empirical (seeming) methods, is the abandonment of critical hermeneutics as if it were inconvenient baggage, an unnecessary impediment to the computational techniques we press into service for our tasks. But why? Without any doubt, computational techniques are themselves hermeneutic, and the possibilities for pushing critical approaches beyond the limits of reading practices formulated within the traditions of twentieth-century philosophy and its paradigm of the hermeneutic circle and into a probabilistic, processual, and constructivist model have never been richer, more possible, and even more pressing.

Ivanhoe: Designing a game of interpretation

In the early 2000s, in the context of SpecLab at the University of Virginia, McGann and I led a team that incorporated many theoretical principles into the design of Ivanhoe.[15] Some of these had already featured in temporal modeling design. Others were implemented in the working prototype. Among these was acknowledging the positionality of historical subjects as intrinsic (integral) to their reading practices. A reading depends on the particularities of the start condition of interpretation. The decision about which features of any text come to the fore will vary from reader to reader, even from reading to reading by the same person.

When McGann and I began conceptualizing Ivanhoe, we were trying to take on this challenge and create a game of interpretation. The structure of Ivanhoe was envisioned to call attention to many of these basic principles. For instance, we insisted that no text was simply self-evident or self-identical. Therefore a text had to be "called"—identified and described, justified and declared, not just "brought in" to the "discourse field" as if it (the text) were an original or self-evident, primary artifact.[16] Its individual features and identity, provenance and bibliographical history, were part of what was explored. The platform was designed to support altering the texts as a deliberate act of reading. By encouraging a transformative reading, we were emphasizing the fluid (and provocative) character of a text as an encoded and probabilistic field into which a reading intervenes. The effect of each reading was visible in the form of a "move" made in the game space. Since we believe that a reading produces a text, we felt that had to be shown. We pushed for the social production of a text to be a line of research. This was manifest through linking and situating an artifact within an open-ended network of related objects and commentary. We required that individual players self-consciously and deliberately identify a role for themselves, situating their work in an acknowledged fiction of historical circumstances. We identified the field of play as always constructed from and seen through an individual perspective. We did not allow an "outside" position to the discourse field. The scene was always seen from a point of view, or, as McGann, following the nineteenth-century poet/painter Dante Gabriel Rossetti, says, "an inner standing point." All the work of interpretation took place in social space, that is, it was part of a game played with others.

Ivanhoe was deliberately conceived as a scene of interpretation.[17] Its features embodied many of the fundamental tenets of our model of interpretation (critical hermeneutics). The "role" situated participants deliberately within a set of conditions and cultural identities. The log of gameplay traced the dynamic interaction among players and the effect of their "moves" on each other. The social space of the game could only be viewed from within, in a situated manner, from the point of view of a player. Knowledge of the game was therefore always partial. In sum, these were parts of a probabilistic model, but the design did not go far enough to demonstrate many of the larger implications of modeling interpretation that are described in chapter 5 of this book and the appendix.

Modeling interpretation: Implications and work ahead

Theories of interpretation have shifted dramatically in the twentieth and twenty-first centuries, as per the opening note about Bruns's characterization of the "Copernican revolution" in critical work. This created a dynamic model of the interpretative (hermeneutic) activity as *understanding*. These approaches are reliant on notions of the materiality of a text—the specific properties of its language, formats, even instantiation in print or digital form, and so on. These properties are subject to transformation, but so are concepts of materiality, which is often reduced to the idea of literal, physical, properties. Recent attention to materiality within a context of new materialisms has posited some agency to material artifacts, rather than continuing the long tradition of seeing them as static, formal, physical things. In addition to formal (organization and structure) and forensic (physically evident) notions of materiality, and the often too-literal approach to media specificity, the performative dimension of materiality needs to be taken into account. This has real implications for the ways artifacts, texts, and objects are remediated into digital environments. The performative conception of materiality is premised on the idea that we do not ask what an artifact *is,* but instead, what it *does.* At the formal level, a work is a set of encoded instructions for reading, viewing, listening, or experiencing. In a performative approach, the cognitive capacities of the reader make the work through an encounter. The humanist inventory of critical methods appropriate to such analysis is long and rich. Textual studies in the traditional and more radical modes, from close reading and new criticism through deconstruction and poststructural play, come into account. Cultural studies has its role here as well, introducing the decentering that shifts the ground from under the certainties of a single worldview, faceting any object of study along lines of inquiry that relativize judgments and values. Exposing the ideological assumptions of digital materialities and the strategies on which they claim and gain cultural authority is essential. The performance of a work provoked by a material substrate is always situated within historical and cultural circumstances and particulars and expresses cultural values (ideology) at every level of production, consumption, implementation, and design. Even the materials are understood within cultural frameworks and signify accordingly.[18]

Engagement with digital media changes when we conceptualize our projects and problems in accord with these critical tenets. Not only is our view of digital objects changed, but we see the possibilities for using the digital environment to take apart the "is-ness" of things. We can shift from an entity-based to an event-based conception of media and demonstrate the radically constitutive, codependent relations of complexity we overlook when we mistake a web of contingencies for a static, fixed, object of intellectual thought. Putting theoretical interpretation into dialogue with digital technology, we engage the opportunity for exposing the very processes by which reification takes place. Again, the ethical issues are glaring and present since these objectifying conditions mask production. We constitute our objects of knowledge through the acts of interpretation that pretend to be observations of what already is. Perversely, the very act of putting humanists into digital projects seemed to bracket critical thinking from the design process (and take design out of the critical process). Here is where the challenge lies—not merely in critical analysis for the benefits of insight, but for the rethinking of design premises. How to bring these conceptions of materiality into the design of digital humanities projects? The answer is not to reinvent humanities theory, or critical epistemology, but to call it back into play in the design process.

As I have said repeatedly, the distinguishing feature of the humanities is a commitment to interpretation as a form of knowing. That knowing is staged as an informed encounter, an event, rather than a reified entity or thing. Though we refer to humanistic *knowledge* (by which we mean familiarity with a corpus of texts, events, personages, beliefs, or epistemological traditions and methodologies of various kinds), the *interpretative act* is always performed anew.

A few basic tenets guide humanistic approaches to interpretation. Each of these builds on and follows from the others, and each is a condensed, summary statement of complex critical histories and legacies. These realms are referenced in parenthesis, with full recognition of the problematic nature of such a list and such reductive references.

The first is that no text is self-identical or transparent; all texts (images, sound, etc.) are already interpretations, there are no originary or original texts (structuralism, poststructuralism, deconstruction). Second, all texts are encoded fields, provocative and probabilistic, into which an interpretation is an intervention; a text is always produced by a reading (semiotics, textual studies, critical theory). Third, any artifact of human expression is a social production, it is a snapshot or time slice through a complex network of social relations (bibliographical studies, critical theory, social history). Fourth, the materiality of an artifact is an index of that network of social relations, where materiality has to be understood as a mediation, a mediating site, in which the material instantiation is not itself to be reified as a set of entities, but conceived as a field of constitutive tensions and relations (media studies, bibliographical studies). Fifth, as a production of social relations, an artifact is a mediating site of power relations that have to be read as a codependent system of exchanges (feminism, queer theory, postcolonial theory, deconstruction, cognitive studies). Finally, we all enter into the act of interpretation as historical subjects, masked from ourselves—we play a role with identities created through cultural conditions and personal fictions; subjectivity is registered as position (social and structural) and inflection (affect and specificity) (psychoanalysis, critical theory, poststructuralism, cognitive studies).[19]

Each of these tenets adds to the framework supporting a probabilistic approach.

Probabilistic aspects—design challenges

Probabilistic and nondeterministic models of interpretation draw on radical constructivist epistemology and materialistic models of the hermeneutic encounter. How might these be modeled in a way that supports innovative visualizations? At the level of encoded protocols (operating systems, machine languages, compilers, programming) computational environments are fundamentally at odds with qualitative approaches. We can interpret these instruments and operations from a critical perspective and also build humanities content on their base, but the logic of the systems is not altered. Alan Turing's early experiment, conducted with the assistance of Christopher Strachey, to produce a Love Letter Generator on the Mark I at Manchester University in 1953–1954, shows just how easy it is to produce apparent nonsense, even while using completely logical methods, in the same way that Lewis Carroll manipulated logic in *Alice in Wonderland* to produce coherent narratives of improbable events.[20] But to incorporate the performative and probabilistic methods of the humanities into computational techniques on a systematic basis—and give them graphical expression—will take effort. The anxiety caused by trying to show ambiguity, to inscribe the situated and partial nature of user-dependent knowledge, works against the realization of these projects.[21]

The humanistic tradition is not a unified monolith, but the intellectual traditions of aesthetics, hermeneutics, and interpretative practices are core to the humanities. The insights of critical theory from a wide variety of perspectives have been brought into discussions of computational techniques, of course, but not with the notion of pushing new designs and construction. The important contributions of code studies, platform studies, and critical digital humanities are background for redesign, but rarely active contributors to its envisioning. The task is to shift from studying the *effects* of technology (reading social media, games, narrative, digital texts, and so forth) to using humanistically informed theory to *design* the technology. This design work has to involve the rethinking of protocols, data structures and format, and information architectures, not merely surface expressions. For instance, the long-sought vision of a pluriverse (in contrast to the "universe" of a single worldview) has yet to take shape in the computational environment for humanities work or in the conventions used for its graphical expression. The notion of a pluriverse was that of multiple views of an object or issue, rather than a single one. The idea was that printed matter was almost always the expression of an individual point of view, but digital critical modes would support an infinite number of facets or perspectives. And yet, no precedents for this kind of approach have been developed. We remain locked into a single point of view, a display on a single flat screen that merely appears as a statement, effacing its mode of address to the viewer.[22]

At stake here, and in all of the projects grounded in these principles, is the claim that humanistic methods are authoritative on their own terms—as demonstrations of partial, situated, and often ambiguous user-centered (hermeneutic) models of knowledge. Modeling gender, as per the examples above, is one demonstration of creating an argument across a richly populated discourse field.[23] But can the model be abstracted from the evidence, given a logical data structure, and compared, on this basis, to that of other models using the same material? The shape of the model has a connection to the structuring and ordering of argument and the way it uses evidence. A project using three hundred documents and citing them in clusters, by chronology in one part of the project, by author or theme in another, and so on through every imaginable attribute gives rise to a schematic diagram. Such evidence might be related to an external reference frame, such as a map that progresses over time, but it could merely etch its configured form against the receptive space of the screen with tokens and surrogates for each node of information, a pathway that shows the argument structure. By itself, this is an expression of a structured connection of evidence. But when contrasted with an alternative, the diagram reveals choices and decisions in sharp relief. Difference and comparison, again, are striking ways to demonstrate identity. Even when the work of such models is done intuitively, using graphical means of making interpretations, the outcome can be assessed quantitatively provided the models of data capture are sufficiently nuanced.

The possibilities are as unlimited as the projects. No single model of evidence and argument will be the same as any other. The weight given to evidence, degree of proximity, frequency of citation—any of these factors can be given a metric value on which to create a stable model for contrast and comparison. Such a model can also be used as a means of navigating and orienting a pathway through evidence, layering and filtering, eliding and slipping, ordering and arranging one component and another. Though the incidental shape that emerges should not necessarily be read as semantic—some features such as proximity or order might have such value but other graphic elements may be an accident of display or convenience for legibility—the form as a whole provides one view of the model of argument. Using the model to show alternative paths, choices not taken, evidence eliminated or added, the whole intellectual palimpsest of process as an emerging field stresses the probabilistic nature of interpretative work with its could-have, might-have, may-still possibilities latent within the discourse field and the work of constituting an interpretation. The ethics of inscribing point of view, location, the place from which an interpretation is made—or a statement of any kind—requires that a means of indicating such ownership be embodied in the models and their display.

Ambiguity and uncertainty

Realist and mechanistic approaches depend upon the idea that phenomena are observer-independent and can be characterized as data. Data pass themselves off as mere descriptions of a priori, or given, conditions. Treating observation if it were *the same* as the phenomena observed collapses the critical distance between the phenomenal (perceivable) world and its interpretation. This undoes the basis of interpretation on which humanistic knowledge production is based. We know this. But as humanists, we have been ready and eager to suspend critical judgment in a rush to visualization. At the very least, humanists beginning to play at the intersection of statistics and graphics ought to take a detour through the substantial discussions of the sociology of knowledge and its developed critique of realist models of data gathering. At best, we need to take on the challenge of developing graphical expressions rooted in and appropriate to interpretative activity. Because realist approaches to visualization assume transparency and equivalence, as if the phenomenal world were self-evident and the apprehension of it a mere mechanical task, they are fundamentally at odds with approaches to humanities scholarship premised on constructivist principles. I would argue that even for realist models, those that presume an observer-independent reality available to description, the methods of presenting ambiguity and uncertainty in more nuanced terms would be useful.

Some significant progress is being made in visualizing uncertainty in data models for geographic information systems, decision making, archaeological research and other domains, and in certain digital humanities platforms. But an important distinction needs to be clear from the outset: the task of representing ambiguity and uncertainty has to be distinguished from a second task—that of using ambiguity and uncertainty as the basis on which a representation is constructed. The difference between putting many kinds of points on a map to show degrees of certainty by shades of color, degrees of crispness, transparency, etc., and creating a map whose basic coordinate grid is constructed as an effect of these ambiguities is profoundly significant. In the first instance, we have a standard map with a nuanced symbol set. In the second, we create a nonstandard map that expresses the constructedness of space. Both rely on rethinking our approach to visualization and the assumptions that underpin it.[24] Ambiguity and uncertainty are assertions of probabilistic conditions for interpretation as knowledge. Giving them graphic conventions is essential in extending the vocabulary of visualizations into critical hermeneutic practice.

The history of knowledge is the history of forms of expression of knowledge, and those forms change. What can be said, expressed, or represented in any era is distinct from that of any other, with all the caveats and reservations that attend to the study of the sequence of human intellectual events, keeping us from any assertion of progress while noting the facts of change and transformation. The historical, critical study of science is as full of discussions of this material as the study of the humanities. Thus the representation of knowledge is as crucial to its cultural force as any other facet of its production. The graphical forms of display that have come to the fore in digital humanities in the last decade are borrowed from a mechanistic approach to realism, and the common conception of data in those forms needs to be completely rethought for humanistic work. To reiterate what I said above, the sheer power of the graphical display of "information visualization" (and its novelty within a humanities community newly enthralled with the toys of data mining and display) seems, paradoxically, to have produced a momentary blindness.

Implications

The goal of defining critical hermeneutics as a probabilistic system is to bring it into dialogue with graphical methods that show the workings of interpretation as a culturally situated process. The simple act of contrast is part of this approach. The work of hermeneutics cannot be codified into a set of tools, only a collection of principles whose values and beliefs guide the design and implementation of environments for this work.

Visualization is a critical aspect of epistemology. Too often, visualization is a representation (much mediated and remediated) passing itself off as a presentation (mere statement). Thus visualizations are often assertions or arguments that pass as declarations or statements when they could be calling attention instead to their rhetorical qualities. A visualization is a representation because of the stages of remove from any phenomenon from which capta were derived. The lifecycle of its production—parameterization, modeling, quantification, sampling, etc. and the processing into formal expression—is concealed. But a caveat here: A representation is a surrogate, a stand-in, which may or may not (need not) have qualities and characteristics of what it represents. On another level, it is not always a representation. No image is able to be equivalent to what it represents. It always has features of its construction, and it serves as a document/expression of a method of encounter between a human consciousness and a phenomenon.

Questions of aesthetics do not go away in these discussions. For one thing, in the history of human culture, some works are more interesting and engaging than others. But aesthetics play a role in the methods being articulated here, since they stand as the very possibility of transformation. Aesthetic imagination offers modes of thinking, seeing, understanding. Aesthetic modes articulate the distinction between directed attention and generative attention, between the didactic and the experiential. The probabilistic model of critical hermeneutics is not a "freeplay" of interpretative work, but a constrained set of possibilities. By their very exceptionalism, aesthetic objects perform important cultural work. McGann, again, states "that a literary work codes a set of instructions for how it should be read. Unlike machine and program codes, however, these codes decipher to an indeterminate number of precise outcomes. They represent exactly what Jarry called 'a science of exceptions'."[25] This apt observation can be supported by the host of variations in interpretation produced by literary works—or by policy documents, or MRIs, or gestures in a public speech when these are subject to interpretation. But data visualizations have depended on impoverished methods to engage with these complex approaches. The sophisticated capacities of eye and image need to be put in the service of knowledge and models of interpretation. Current practices of reductive graphs and standard charts have debased interpretation, made it simplistic and literal. We know better and can do more.

In a nonrepresentational approach, which will form the focus of the next chapter, visualization functions as a space for interpretation and for the act of making an argument. This approach shifts attention onto the argument structures. If current visualizations are inadequate, which they are, it is partly because the models they present are linear, mechanistic, based in empirical methods. But interpretation is nonlinear, probabilistic, emergent, and stochastic. This is what makes the link to politics through an approach to cultural/social processes that is based in similar insights about cultural phenomena. Cultural phenomena are not linear. Politics are driven by affective forces, not reason. Reason, as a construct, emerges from Enlightenment thinking that was wedded to empirical methods. Reason provided the foundation of the modern democratic project—and of Newton's classical physics. But Reason, especially in reified form, was only a construct, not an operative force. Calculating the probabilistic relation of understanding to the unfolding conditions of the real (however imagined) will require better methods of visualization and analysis. This work has to be informed by probabilistic hermeneutic methods if its complexities and contradictions are to be engaged and supported. The tools we develop for analysis of the cultural record will work in the present as well as for materials of the past. If ever there were a moment when we are in need of humanistic methods to expose the workings of ideology, it is now, when the forms of mediated expression work through the social imaginary with potent force using computational techniques and visual methods as instruments of authority. We always need to be attentive to whose interests are served by these techniques and their capacity to conceal the processes by which they work. Visualizations conceived within a humanistic framework may have potential in the social and political spheres as well as the literary ones.

3 Graphic Arguments: Nonrepresentational Approaches to Modeling Interpretation

Arguments about nonrepresentational approaches to graphic activity weave throughout this book for a reason. The term *nonrepresentational* is not meant to suggest that a work is not visual, but instead, to suggest that it is a graphic made as a primary act of knowledge production. Nonrepresentational images are ones that do not serve as surrogates for an already existing object, whether that object is a thing, a place, an experience in the world, or a data set. In this context, the term nonrepresentational emphasizes the constructed character of knowledge production as interpretation. Though there is no barrier to creating nonrepresentational graphics from an empirical or positivist perspective, the use of the term within Nigel Thrift's work in geography (which I take as my source) was meant to emphasize experiential models of space presentation. These were counterposed to conventions of mapping that assumed space as a given, with a map as its surrogate, in a stable, largely unproblematic relation between the two.

Understanding the distinction between visualization and modeling here is essential. Information design is a subset of graphic design. In the words of one the foremost practitioners of visualization, Edward Tufte, the task of the information designer is to "show the data" and to "avoid distorting what the data have to say."[1] Tufte also goes on to say, "Graphics reveal data." Tufte's conviction that information exists independent of—or in advance of—the presentation of data in graphical form is problematic. A spread sheet may hold "data," but his idea is that the data have a "form" that can be "revealed." On a pragmatic level, information designers understand their task as the creation of clear, legible, unambiguous presentations of this data. They see "data" as quantitative values stored in an analogue or digital format and the role of the visualization is to show patterns in this information. But every graphic representation is also a rhetorical device.

Every presentation structures arguments—it doesn't simply "reveal" facts, or forms, in all their purity. The relations between *what* is communicated and *how* have to be acknowledged. A spread sheet is an embodied format. It has particular graphical properties that structure our reading of the values it holds. Even such a basic visual format can be understood as an argument structure that communicates through graphical organization.

Formats are semantic by virtue of their structure. The column and row organization of a spread sheet produces meaningful relations. Charts and graphs carry values through labelling, data storage, and sometimes imagery that links them to specific domains. The conventions for visual display of knowledge are *performative* (they make and generate knowledge) and *representational* (they can be put at the service of display of apparently preexisting knowledge derived from a table, chart, or text or embodied in an object in the world).

The inventory of types of visual information structures isn't long: bar diagrams, trees, maps, pictorial diagrams and icons, flow charts, bubble charts, and tables. Taking each of these in turn we see that:

- Bar diagrams derive from statistical analysis and function through supposedly unambiguous distinctions expressed in a grid of rationalized information readily available for comparison according to a standard metric.

- Tree structures, by contrast, derive from genealogy and evolutionary biology and suggest continuity of dependence and kinship in the flow of information across generations.

- Maps are the record of explorations, phatic and tactile, narrative and immersive when created from inside the experience of discovery, but rationalized through projection when produced from outside, as images. The complexities of representing a curved form on a flat surface, as well as the many cultural imperialisms at work, provide their own history within the range of projection methods.

- Pictorial diagrams that make use of icons and other images have a cultural specificity through their particular details, the way they are rendered, as well as the inventory of images used (to show a woman, for instance). Pictorial images carry and evoke multiple associations.

- Flow charts have their origins in management and organizational structures. The directional force of power relations and movement of goods through a production system often has a conspicuous absence of human agents, as if processes were an inevitable and natural fact.

- Tables and grids work by putting discrete cells of information into meaningful syntactic relations with each other—a classic example is the timetable.

- Network diagrams assume an absolute distinction between nodes and the edges that connect them, and their display is organized by algorithms that optimize display rather than following strict interpretation of the relationships among the nodes. Their visual arrangement cannot be read as an accurate presentation of information, only an approximation.

These forms are usually displayed on a single, unified and rationalized surface—page or screen. The vocabulary of charts and graphs provides only a very elementary starter set of graphic forms structured on linear, hierarchical, and tabular models. More complex spatialized and temporalized relations can also be described through topological (mathematical) approaches that maintain basic logical relationships in spite of distortions and changes in scale. Topology emerged from the eighteenth-century mathematician Leonhard Euler's struggles with the Königsberg Bridge problem. In 1736 Euler established what he called a "geometry of position, not of measure" as a foundational principle.[2] Johann Benedict Listing, in 1847, was the first to use the word topology to discuss the connectivity of surfaces.[3] While used in modeling events in chaos and complexity, topological models have rarely been put in service to humanistic interpretation or its presentation in graphical form. Given the complex systems involved in narrative, discourse analysis, and event structures, it would make sense to make more use of these and other mathematically sophisticated models.

Graphical input

When Ivan Sutherland created Sketchpad in 1963, he engineered a system of direct graphical input from user manipulation on the screen to data storage.[4] The images he created were not based on data, or even on formulae or algorithms, but created as free expressions of his light pen registering in pixels. Interestingly, the system was designed to store the information as a topological form, keeping the logical structure of the graphic intact in spite of morphs or changes in size, scale, position, or orientation.[5] This approach to visual information lacks the precision and specificity of inscriptional approaches, but still demonstrates the primary point—that images can *create data sets and structures* rather than simply *express them*.

To relate the technique of graphical expression to scholarly interpretation requires thinking about the basic components of concept modeling (a term used in data design to indicate a way of making a model of an idea, field, or problem) through visual means. What are the semantics of graphics, and how can argument structures be made legible through visible methods?

In the sciences, visual modeling techniques are frequently used as a way to test a thesis or create new knowledge through experiments that bring an idea into form. Visual modeling techniques and procedures are rarely used in humanistic approaches to interpretation. In part this is because our work tends to be text-based, and determining what is to be "quantized" in a text is sometimes difficult. The rise of information design and increased familiarity with display techniques as part of computational analysis in digital humanities has softened the resistance to visual presentation of data. But approaches to interpretation that use visual modeling as a primary method of analysis—to create the data (especially structured data), not just display it—are less familiar. If a narrative structure involves multiple figures and agents, it will need to be charted in ways that involve emergent properties, not merely linear chartings of its events. The intuitions that govern interpretation are not governed by mathematical systems, but their graphical expression can be processed into a data structure.

How can we understand the relation of representation and model? All visual images are expressions of models, if by model we mean an abstract, schematic structure that expresses a concept—but not all visual images are *only* models. Many are representational, or decorative, or contain all kinds of accidental and incidental information. Representations appear to be self-evident, but we can qualify this by showing that a representation is a special type of model. Representations exhibit different degrees of isomorphic (structural) connection between the visual image and its referent. Charles Peirce's categories of icon, index, and symbol define three basic relations of resemblance, continuity, and arbitrariness. But even the most visually analogous image (e.g., a realistic rendering of a frog) is an expression of a model (in this case, it expresses the concept of visual analogy and isomorphism or structural similarity).

Even the earliest graphical expressions of human activity demonstrate basic principles that remain operative in current forms. A Neolithic calendar found in Bulgaria and dated to the 4th or 5th millennium BCE uses a grid to organize the marks tracking the phases of the moon.[6] A high degree of abstraction is required to create a structured system to conceive of temporal events in such elaborate relations. Calendars are visual forms designed for use, not just static display like, say, religious icons. Calendars are surfaces organized to put temporal units into relations with each other in a very specific way. The surface has been rationalized, organized by a system of coordinates that structures information so that we read it relationally. But in fact, the abstract structure isn't a representation of time, it is a model of time that allows for calculation or computation. The structure and the information are not identical with each other. The calendar models temporal elements so they can be manipulated and managed (organized into days, weeks, and months with repeating patterns). It doesn't just represent some "natural" condition of time. By providing an idea of the way time is structured, it embodies this in the graphical presentation, but the schematic abstraction allows different combinatory possibilities to be produced. The calendar grid suggests that the days have both an ordered sequence and are part of grouped and repeating events—ideas that are readily apparent in the visible evidence. Similar abstract capabilities are evident in the tokens devised in the earliest counting and accounting systems. The principles that organize cuneiform writing in the 3rd millennium BCE make use of numbers, symbols, and word representations simultaneously. A mark doesn't represent some *thing* in such a system, it represents an instance of a type of thing within categories. The classification system for this might not be rendered explicit, but it is nonetheless fully operative, and legible, in the graphical code.

The distinction between model and representation allows for self-consciousness about the rhetoric of graphical expressions. Unlike representations (which are surrogates), models have the capacity to generate new intellectual insight, not simply represent what is already known in a graphical form. A bit more clarification of the basic distinction between the idea of modeling knowledge and representing it may be useful. Representations are always premised on abstract conceptual schemes—or models—that shape any individual expression within constraints and patterns of thought. But a representation is not a model, it is an image that embodies a model or models but gives them a specific expression. Though our ideas of what something should be—a house, an airplane, an automobile—constrains our ability to design these things within an abstract model, once we make the house or plane, it is also an expression of a model. Breakthroughs in knowledge come from changing the model, or by innovative expressions.

Changing uses of computer-aided design have signaled a shift in attitudes toward architecture and the manipulation of forms through abstract, graphic processes. Of course, a change in a representation—making a three-legged frog or a monkey with antennae—might call forth a new model, but it is still serving as a re-presentation of that visual form or idea.

As already noted, the term nonrepresentational does not mean that the approach contains no graphic features. Quite the contrary, nonrepresentational approaches use graphical means as a primary method of modeling human-authored interpretation rather than displaying preexisting data sets, as we shall see.

If I draw a line, a single line, on a piece of paper, across a bit of territory, between members of a group, it divides one area of paper, space, group, and gathering from another. The line does not *represent* a division, it *performs* that division. Neither the identity nor the function of that line has anything to do with imitation. The line does not resemble anything beyond itself. It does not stand for something else, even if it can carry all kinds of values and resonances and become a border, an edge, a defining demarcation of exclusion. The line is the quintessence of a performative graphic expression. The line is visual—sketched, etched, drawn, or merely traced—it is palpable and tractable as a line, as a graphical trace, an expression, a thing that has dimension, duration, extension, all the properties of form. It signifies first by virtue of having been drawn. The coming into existence of the line performs its primary work. But it gathers signification, or meaning value, through the resonant features of its place and position, the conditions within which it registers as a line, and by the attributes it carries—thin, thick, clean, rough, straight, mechanical, etc. The list of attributes can be extended infinitely, and they may carry and/or call forth meaning through associations. But the primary act of making a line as a single inscription is fundamentally and profoundly nonmimetic. The drawing of the line is performative. All nonrepresentational graphical expressions are performative and nonmimetic in the same sense. They are generative. They are primary means of creating signification through graphical methods and expressions. Even if they "look like" something else, they were generated first and then assumed their air of resemblance afterward. The importance of this for modeling interpretation cannot be overstated.

Rationale for nonrepresentational approaches

This section presents an epistemological rationale, intellectual justification, and design outline for a nonrepresentational approach to modeling interpretation in a graphical environment.[7] It begins with a critical discussion of the representational approaches that are the common form of information visualizations and suggests that the less familiar nonrepresentational approach could be used to augment these existing visualizations by supporting interpretative work that is closer to the practice of humanistic hermeneutic traditions. Representational display, based on large-scale processing, surrogates, and conventional visualizations, and nonrepresentational modeling at the level of the individual interpretative act operate at very different scales and support distinct intellectual work. In a representational approach, data precede display. Display is a surrogate produced according to automated protocols and algorithms. These cannot be altered or intervened except through rewriting their code, and the display, though interpretative and subject to interpretation, cannot be used as a means by which interpretation is actually modeled. While all visualizations express a model, they do not all provide a modeling environment. In the nonrepresentational approach proposed here, graphical input serves as a primary means of interpretative work. More significantly, a graphical environment that supports direct modeling of interpretation allows traditional humanistic approaches, close reading and marking of texts, documents, artifacts, or images to be integrated with computationally produced visualizations.

While information visualization has become ubiquitous in digital humanities, common tools for graphic expressions of data have limited value as primary modes of creating interpretation. They do not provide an environment in which interpretation is actually done directly. The traditional work of scholarly interpretation, at the level of individual artifact or text, often seems at odds with the computational processing that produces data visualizations. An alternative, grounded in a nonrepresentational approach to modeling interpretation in a graphical environment, could add essential capacity to the existing methods and platforms by providing a space for direct creation and inscription of interpretative work.

User-authored interpretation is premised on the distinction between display and modeling and signals a crucial difference between the two approaches. The modeling approach uses graphical means to produce interpretative work using visual argument structures such as contradiction, ambiguity, parallax, and point of view that are fundamentally interpretative in character. It engages conventions and dimensions of graphicality not used in standard chart, graph, timeline, and mapping software with their entities and attributes. The system I developed in the context of the 3DH project at the University of Hamburg is meant to be a radically innovative addition to existing visualizations and to add new dimensions in the service of interpretation and modeling alongside representation and display. The term *modeling* is being used here to refer to the process of creating a generalized intellectual schema or structure and should not be confused with the way the word is used to refer to three-dimensional rendering software.[8]

As noted here repeatedly, the visualizations adopted by digital humanists (charts, graphs, diagrams, maps, and timelines) were mainly developed in the natural sciences, social sciences, statistics, business applications, and other fields. These bear the hallmarks of positivist, quantitative and/or statistical approaches to knowledge that limit their application to interpretative practices in the humanities. This limit is structural and procedural: the work of interpretation cannot be readily performed through the display, and the algorithms and protocols cannot be altered through manipulation of the display. The display algorithms and protocols that generate visualizations are human authored, and thus perform interpretative work, but they are sealed off from direct engagement. Even when displays support a filtered, faceted search for discovery, they are not a means of inputting or transforming data or of modeling the interpretative work. Practitioners skilled in statistically driven work are keenly aware of the ways data production embodies interpretative decisions about the statistical complexity and lifecycles of their data (from parameterization to display). But the conventions they use in these visualizations remain linked to a representational paradigm.

In a representational paradigm, the relation between data and display is unidirectional, the data precedes the display, and the data is presumed to have some reliable representational relation to the phenomena from which it has been abstracted. The display functions as a surrogate for the data—which is itself a surrogate, adequate or inadequate, for some phenomena. Simply put, the display stands for the data, is a re-presentation of the data. But, as has already been stated, visualizations are generally taken to be a presentation, a statement (of fact, or argument, or process), rather than a representation (surrogate) produced by a complex process. Because of their presentational appearance, visualizations, as has also been stated above, are what we would term *declarative* statements. In the declarative mode (in contrast to the interrogative, the conditional, the subjunctive, for instance), statements are not qualified; they seem to simply state *what is*. As a result, the lifecycle of data production is concealed in these visualizations; the features of their display (proximity, size, scale, color, etc.) are often read as semantically meaningful when they are frequently, actually, the result of display algorithms optimizing screen space, legibility, or other factors that are not intrinsically semantic. As conspicuous aspects of the display, these structures and graphical features are often taken as significant (how close something is to something else, for instance, might be simply an artifact of the display that gets read as meaningful). Reading the artifacts as if they were the underlying or original phenomena, or even accurate representations of or surrogates for it, ignores the complex processes of production and construction. Some features of visual display are semantically meaningful, others are not.

Instead, we should consider that visualizations are usually representations (constructions) passing themselves off as presentations (statements of self-evident fact). Again, in the representational mode, visualizations usually lack certain features. They do not carry author attribution, nor do they contain any account of the lifecycle or parameterization of the underlying data. They are generally produced from a single viewpoint. Finally, they are impervious to direct input or variation in real time. They also erase the circumstances of what we might term their enunciation, or articulation, and thus, they do not contain any markers of the historical and cultural conditions of their production. The ethical issues in assumption of such value-neutral visualizations are similar to those in any other human expression, leaving them open to the same critiques of unacknowledged bias. When they are dynamic, visualizations use conventions based on an overview and zoom model, which supports faceted search, detailed query, and filtered display. These display methods can be very useful when applied to humanities projects and research. In their capacity as discovery tools, these visualizations use graphical display to expose patterns in data, to see data, and ask questions of it. This is particularly valuable for large data sets, and this line of argumentation has been the standard support for the use of visualizations across fields. However, the display and discovery method of visualization does not exhaust the possibilities for the development of other graphically enabled interpretative work. Furthermore, the classic formulation of discovery is based on a consumerist model of user experience, rather than one of generative engagement and critical dialogue.

Concept modeling differs from standard visualizations in nearly every approach to presenting information or interpretation in graphical form. Locating concept modeling in a succinct typology of visualizations should help make the specificity of the project clear. Visualization software can be divided into the following categories: (1) drawing programs that generate images algorithmically (e.g., Processing) or through rendering (e.g., Rhino) in pixel/raster or vector formats with surface textures and other visual effects; (2) visual displays of quantitative (numerical or statistical) information (Tableau, Google Fusion, Excel charts, scatter plots, etc.); (3) forced or directed graphs (Cytoscape, Gephi, and other network visualizations) generated through computational analysis of betweenness and other factors; (4) simulations of complex, nonlinear, or dynamic systems (Game of Life, VisSim); (5) visualizations from integrated data analysis (Inscriptifact, imaging, forensics, etc.); (6) visual presentations of data mining or analysis (Voyant, Google Fusion Tables, Tableau). None of these are experience-based, all are driven by strict quantitative and/or probabilistic statistical or algorithmic methods for analysis and display. All use standard metrics, uniformly measured spatial/graphical display, and continuous spatial environments. While these features are adequate for the visualization of information conceived within terms of homogeneous metrics, they are not adequate for the creation of models of varying, discontinuous, or inflected experience of temporal, spatial, textual, emotional, or affective phenomena. While humanistic documents and materials can be reduced to quantitative data through certain processes of abstraction, and for particular purposes (counting and sorting), the dimensions of hermeneutic thought that play a major role in many humanistic works and enquiries cannot be modeled in graphical systems based on features extracted from the natural sciences and its empirical methods.

By contrast to standard visualization approaches, concept modeling does not assume the existence of data or other representations in advance of the act of interpretative work. Modeling is a primary, productive, interpretative act that can be used to create data. Modeling does not re-present data in a chain of surrogates (from parameterization / extraction or abstraction / reduction / standardization / and presentation). Instead, a modeling environment consists of graphical components, activators, and dimensions culled from visual and pictorial traditions to be put at the service of high-level concepts (contradiction, ambiguity, comparison, etc.). Modeling assumes that graphical platforms can support interpretation as primary means of knowledge production and/or interpretation.

A simple example should help make this clear: if I have a diagram on my screen and decide that two of the points in the display are related to each other in a particular way, I draw a bold line of connection between them. Connection is an interpretative concept. Connection is not a thing, not an entity being represented, it is a concept that is being modeled. The two points may have been part of a representational display, a conventional chart or graph. They might also have been created on a blank canvas or placed on a map or a timeline. But the point is that the connection between them is expressed as a deliberate and direct interpretative act that is performed graphically by drawing the line of connection. The line is used in the service of the concept and inscribes the interpretative model on the screen. This action models interpretation about the information in the visualization, using graphical means. The existence and weight of the connecting line are then registered in a table or other data structure or format. The data structure can hold a simple quantitative value or an expression of value, calculated as a factor of a variable that changes over time, or is calculated to any level of complexity.[9]

Concept modeling thus supports direct acts of interpretation in the graphical environment. The visualizations are models of a particular interpretation, and they bear the signs of their production in author attributes, interpretative layers, and other features that stress the "spoken" or articulated (enunciative) aspects of a graphical visualization. They are clearly and markedly rhetorical. The term nonrepresentational, as it is being used here, is borrowed from work in critical cartography and nonrepresentational geography. In that context, the term nonrepresentational is used to suggest that a map does not precede experience or a phenomenological engagement with landscape and its features, but is instead made as an inscription of experience.[10] The presumption of representation as an adequate surrogate, as knowledge, is therefore countered by the assertion that knowledge precedes inscription and presentation. The sequence of epistemological events is reversed. A map does not represent a territory, but is an interpretative and discursive artifact.

In concept modeling, the two-way potential of the screen (or other input field, if the interface is created spatially or in physical computing modes or other alternative modes) is activated, and the screen serves as a primary site of work. Interpretation is enacted in the screen or platform. The concept modeling environment is designed to include elements, activators, and dimensional features that can be put at the service of interpretative work. These features function at a high level of conceptual work, such as the example of connection given above. By contrast, representational graphic platforms generally consist of a specific set of entities and attributes meant to represent data or things (e.g., timelines consist of points, intervals, dates; maps might have cities, rivers, roads, or borders; and so on). The graphical features of the concept modeling environment are designed to express fundamental principles of interpretation: uncertainty, parallax, contradiction, partial knowledge, and so on. For instance, instead of a timeline that represents events as points, a temporal model is constituted by relations of before and after, simultaneity, duration, slow and fast time spans, and variable models of historical chronology or other conceptual elements.

Concept modeling creates data input through direct manipulation of graphical features on a screen. Imagine a line has been charted from values in a spread sheet. Now imagine that a line of connection has been drawn, a graphical inscription of an interpretative action. The table on the spread sheet can register the addition of a new set of attributes. Extra attributes, termed "affective" to indicate that they are generated through human, individual decisions about significance and importance, can be added. These might require complex calculations of their value if they register in the table as forms that designate forces, probabilities, and other dynamic or emergent properties. No limit exists on the type or value of additional attributes or the relationships that might be modeled in the graphical environment by drawing directly on the chart. The modeling would simply continue to change the underlying data structure through a direct act of interpretation inscribed using graphical means.

Justification

The justification for nonrepresentational approaches to modeling interpretation is that visualizations generated by display protocols have served very well for large data sets processed according to statistical methods, and though these are human-authored and therefore already interpretative and rhetorical, they produce results in the form of statements that conceal or largely ignore these aspects of their presentation. The traditional approach to hermeneutic analysis in the humanities is modeled on the idea of a generative reading of a text, event score, performance, or other artifact of the cultural record. Such reading is generally close, individual, and meant to create an argument about how the artifact or text under consideration can be understood. Creating an environment in which to enact this interpretative work, generate data from it, and make it an intuitive mode of input is one justification for this project. But the deeper justifications are related to convictions about knowledge—they are epistemological. As mentioned above, the declarative mode of representational visualizations often causes the artifacts of display to be taken as semantic features of the data, or, worse, of the phenomena for which the visualization and data stand in surrogate relation. In addition, the ability to alter the data structure of a visualization through direct input is largely foreclosed in conventional display.

A modeling environment adds the capacity to engage directly, graphically, in interpretative work on visualizations, artifacts, documents, texts, images, or any other file that is being displayed. In this environment acts of specific, authored, varied, contradictory, collective, and other interpretative readings can be modeled, marked, stored, made visible, revisited, revised, exported, and analyzed. The goal is to support rhetorical, argument-based, interpretative work, not statements that appear to be singular or self-evident. Concept modeling is an alternative to (and augmentation of) the declarative mode of conventional visualizations. One of its crucial epistemological premises is to emphasize a subject-centered, or enunciative, system of graphical expression. Enunciation marks visualizations as situated, partial, historical, authored, observer-dependent, and rhetorical. These shifts, from the declarative to the rhetorical, the display mode to the modeling one, the automatic to the hermeneutic, and the flat, sealed, space of the screen to the multidimensional one of virtual intellectual space and direct input are the core contributions of this project. Concept modeling supports humanistic methods of interpretation as specific ways of working, thinking, and producing knowledge.

To summarize, the intellectual principles of concept modeling are a nonrepresentational approach that creates knowledge and interpretation directly, rather than using surrogates to generate display. The distinction between representation and modeling can be understood as that between the activity of "designing" a house and "modeling" a dwelling. In the first

instance, you might begin by laying out a plan with a living room, din-
ing room, front door, bedrooms, and so on that assumes a house contains
"rooms" which serve specific functions: eating, living, sleeping. The design
platform might contain a kit of windows, walls, doors, and other entities
from which the "house" is composed. As an alternative, consider a modeling platform that is composed of high-level concepts: shelter, boundaries,
ingress and egress, scale, pathways, sightlines, degrees of privacy and proximity, and so on. The "model" created as a result is not entity-driven, but
concept-driven, and the graphical platform is the environment in which
the modeling takes place. The platform does not represent a house, it models a dwelling. In the first approach, the entities are set in advance in a
menu of options as a pick list; in the second, concepts model a space whose
functions and specific qualities emerge.

Concept modeling is meant to support interpretative approaches to
knowledge which assume that knowledge is partial, situated, constructed, and
authored (this differentiates it from empirical and positivist approaches whose
assumptions are that knowledge is stable, repeatable, universal, and complete). This distinction can be articulated as the difference between observer-
dependent and observer-independent approaches to knowledge production.

The observer-dependent and observer-independent distinction also
underpins the difference between hermeneutic and mechanistic approaches
to visualization. In hermeneutic visualization, the graphical environment
is the primary space for creating, marking, and processing interpretative
work; in a mechanistic visualization, a display is generated through direct
processing of data, statistical information, text, or other computationally
tractable information (information that can be computed). The display in
a mechanistic visualization stands in a stable relation of underlying data to
representation; the display is a surrogate, even if it can be queried, redrawn,
filtered, or faceted, the display rarely changes the underlying data, and
when it does, it does so mainly as a change in values, not a change in
architecture or structure. In a hermeneutic visualization, no data or other
intellectual information has to exist in advance of the process of graphical
production (though base images such as maps or texts might be used as the
ground on which to register an interpretative, or hermeneutic, production).
The interpretative approach is linked to a deliberately marked system of
subject-centered enunciation or articulation that shows the position of a
speaker and spoken subject.

Design of the concept modeling environment

The design guidelines for the graphical production environment are meant to embody the intellectual principles and goals of the project. This work was originally undertaken in the context of the 3DH workshop at the University of Hamburg in 2016. The designs were meant to address the need for a way to work with the elaborate textual markup schemes created in a platform called CATMA. The goal was for these tools to work with Voyant, as well, a platform for basic text analysis. These two display modes were to be complemented by an argument space that would use the same conventions, but as a primary means of interpretative work along the lines described throughout this book. The 3DH environment was supposed to be able to (a) work with existing visualizations and (b) create completely new visualizations that generate data structures. The section that follows was written as a prospective design brief, hence its tense and tone.

Design brief

The conventions used by CATMA and Voyant are both examples of visualizations generated by platforms created specifically for humanities research. The markup in CATMA is a deliberate act of textual interpretation, and Voyant is interpretative by default of the design of the text mining tools. CATMA displays are already direct and nonrepresentational by virtue of the way they are created through direct input. Voyant displays are representations of text analysis, and thus serve as examples of conventional visualization. Both seemed likely to benefit from innovation.

The new visualizations were designed to either make use of a base image or be generated from a blank canvas. A historical map might serve as a base for one of the examples and provide the foundation on which to create spatial interpretations of historical events. But the creation of relative chronologies with nonstandard metrics was an example of modeling directly in the graphical environment, without a base image.

As noted above, the nonrepresentational approach consists of a generalized set of graphical features, interpretative activators or inflectors (activators and inflectors were terms we developed specifically for the project, with activators as tools that did things and inflectors as graphic attributes that showed values). These features were used to assign qualitative or affective value through graphical inflection (e.g., salience, reliability, etc.). It was

envisioned that any and all of these could be customized for specific projects through labeling or selective use. The dimensions have structuring and
syntactic implications for the relations among aspects of an interpretation,
but they are not defined semantically the way, for instance, the elements
in a map legend might be. The dimensional features are therefore to be
understood as components of interpretative work, not as entities that are
being represented in graphical form. For example, making a comparison is
a fundamental interpretative act, in which values within areas of an image
or text are being put into relation to each other. An act of comparison is
not entity-driven, but process-driven. The dimensions of the conceptual
modeling environment embody principles of such interpretative activity.

The modeling environment was designed to support data inflection, the
process of going from graphical input to data/infrastructure modification.
As a model is created, modified, and augmented, the data structure is generated through a process of mathematical calculation. Regular tables and
standard metrics do not have to be the only data structure created in the
process. Branching, layered, crazy-quilt tables and other anomalies could
also be generated by the conceptual modeling, which registers the many
variable attributes of these structures.

The modeling environment was envisioned to consist of three graphical
components:

- Graphical features (the palette of elements with which to work, drawn
  from standard literature in the field),
- Activators/inflectors (application of the features to enact, inflect, inscribe,
  and mark interpretative moves), and
- Dimensions (these draw on conventions of pictorial form such perspective, layering, parallax, etc. put at the service of the many aspects of a
  hermeneutic analysis).

Graphical features

The basic palette of graphical elements for static and dynamic visualization is shared by representational/display and modeling/interpretative
approaches. The graphic primitives are taken from Jacques Bertin and
Leland Wilkinson and augmented by elements from animation.[11] The
graphic primitives include: tone, saturation, color, transparency, texture,
shape, orientation, size, and position. The dynamic elements include:

order, sequence, twist, decrease, increase, and flip. Other dynamic graphic attributes include torque, weight, force, and attraction/repulsion. None of these carry semantic value in themselves (though, obviously, certain colors and shapes call forth associations, as do contrasts of size, and so on). But all of these elements could be put at the service of concepts, values, or meaning-producing signs in a standard legend.

Activators and inflectors

Among the activators and inflectors were many that expressed *affective*, emotionally charged values. Affective (emotionally charged) attributes are not simple statements of quantitative value but are expressions of qualitative value. For example, when creating a map of a particular spatial experience, affective attributes might be used to register smell, comfort, pleasantness, cold, fear, anxiety, etc., such that the geospatial information is generated as "spatial coordinate plus a factor of X." Such a formulation might also contain a formula for change over time, or modification in relation to any specifiable condition or aspect of the phenomenon. Affective attributes, like other generative and relative metrics, do not have to be arbitrary, but they necessarily include the point of view of their author. The activators introduce syntactic, relational attributes while the inflectors, generally, introduce semantic attributes. Activators and inflectors are not entities, but qualities, and are made of the basic graphical features. Thus, salience might be indicated by glow or luminosity, ambiguity by tonal value and vague boundaries, contradiction by lines of force and so on. Establishing conventions for a set of activators and inflectors relevant to a particular project makes sense, but a fundamental set of argument structures and rhetorical moves, such as those just mentioned, makes sense as well.

Dimensions of interpretation

The concept modeling dimensions listed here are not meant to be definitive or exhaustive. They were the outcome of research for the 3DH project. But they comprise the fundamental moves that can add dimensions to a flat screen space. The system is extensible and customizable, though establishing some conventions of use will lend legibility to the project overall. The dimensions are literal interventions in and manipulations of the screen space put in the service of conceptual modeling. A projection, for instance, should be understood literally and metaphorically. What kind of shadow or form is cast by a data visualization when it intersects with another plane (a value plane, ideological plane, hegemonic system, etc.). Some of these moves, like tilt, for instance, may seem obscure at first, on account of their unfamiliarity, but they are meant to suggest ways of turning interpretative work into systematic metrics so that graphical displays are generated affectively as well as objectively. The metrics are dependent on interpretative values, not mechanistic ones. The model makes the values, and the values become the basis of a system.

A list of the spatial moves envisioned in the 3DH platform follows here, and each is described in turn, though a fuller understanding of these requires the graphical images in the appendix. Some of these are variations on familiar approaches, such as *annotation*. Others are borrowed from conventions of drawing or graphics, such as *point of view* or *tilt*. But some are novel to the point of being difficult to grasp in the abstract, such as *slicing*, though they assume a rich data set that can be exposed along a particular axis or array.

1. *Point of view* is indicated by a number of features such as author attribute, vanishing point, horizon line, picture plane, now sliders (indicating the position of "now" in a timeline), and multiple viewpoints. Point of view embodies the place from which an image is constructed, and thus assigns the image and its scale to a particular, rather than a general, owner or author. Introducing point of view within data displays dimensionalizes them in ways that radically alter the neutral or objective approach to visualization and returns them to an enunciative system. This shift is crucial for moving from a user-independent model of visualization to a user-dependent one. A point of view can be constructed from a historical or spatial vantage point as well as one owned by an individual or a group. Relative scales (see 8. Relative scales, below) can be used to contrast points of view based on differing metrics. Perspective systems embody point of view by using eye lines, vanishing points, and horizons, all of which inscribe scale and positionality within the scene. Single-point-of-view systems are monocular and omniscient, but the use of multiple perspectival systems is also possible to introduce relative viewpoints (see 10. Parallax below).

2. *Layers* allow information to be brought forward, pushed backward, and changed in tonal value or intensity. Layers are used to distinguish a

base layer from the interpretative materials, or, without a base layer, to create a model from scratch. Layers can be used to hold contradictory arguments, varied uses of evidence, or to display any aspect of a project that can be articulated independently.

3. *Slicing* makes cuts across data objects expressed as visual models, graphical artifacts, or displays to reveal patterns across another axis. Slicing is primarily a discovery tool, but it can be used in an active, modeling, mode to create and study patterns, record them, create analyses in an iterative process of modeling and display. Introducing a slice into a model implies depth that can be actively intervened and engaged within the presentational field.

4. *Annotation* adds information, labeling, and/or commentary into any model and can be added to any feature of a data set present in a display: a node, edge, point, text, image. Annotations can be recorded in a data structure as attributes noting connections, relations, or other analytic and interpretative features.

5. *Tilt* moves layers through an angle of interpretative inflection to distort the display. The angle might be generated systematically through a generative metric (rate of change of any variable) or through an interpretative principle (bias, sentiment) or affective metric. The tilt angle can be calculated through a generative process (see 7. Generative metrics, below) from some other feature of the model, or it can be used as a graphical tool of manipulation that reveals some aspect of the model to which a value is then assigned.

6. *Projections* are common modes of creating mathematical transformations in geometric renderings or spatial constructions. Casting a shadow is a basic projection move and creates a new, distorted, derived version of a form through an angle of transformation. This angle, like other affective or interpretative metrics, can be generated systematically or arbitrarily. If I want to show the influence cast on a scene in a narrative by a specific character, I can project it, for instance, according to an angle of influence, force, power, etc., generated through a contrast of power terms, vocabulary, or any other text-mining tool. Anamorphic projections can also be created using generative metrics (see 7. Generative metrics, below). The plane onto which a projection is made can be a mere convenience, a device to hold the projection. But a plane can

also have a defined identity, such as a plane of history, of ideology, of bias, and its angle to the projection can be assigned an attribute (e.g., salience, contradiction, ambiguity, etc.).

7. *Generative metrics* are created by taking a feature of a graphical artifact and using it to generate a new scale. For instance, the lengths of lines between data points on a graph, though they are the result of a change in value, can be used to make a new scale by rotating the line lengths downward to create points on a line that define new metric standard. The scale generated in this way is not arbitrary, but derivative. Such a metric might inscribe biases recorded as biases, and by mapping a graphical element from one metric system to another the distorting effects become evident.

8. *Relative scales* are essential for showing differences among taxonomies and other systems of classification, knowledge production, and epistemological models. Relative metrics can be generated from textual interpretation, intuitive graphing, and other ways of engaging with nonstandardized metrics such as varied chronological scales and relative models of historical time. They can be correlated at certain points of alignment, but never fully reconciled since their units, scales, and assumptions about the completeness of their own models are each distinct. The use of relative scales is essential for exposing the differences among taxonomies, worldviews, value systems, and other features of knowledge production, classification, and management. It is fundamental to the humanistic critique of positivist and empirical models of data visualization.

9. *Fold* allows different areas of a model to be brought into connection with each other in order to see how they relate, what patterns they share and/or make, and what comparisons emerge in the process. Folding is a powerful structuring tool, and making concept models that are multidimensional depends on folds, edges, planes, angles, and other geometric principles. The line of fold can be derived from the action of folding, or it can be determined through a deliberate choice. Folds are able to show change as a contrast of values, rather than as a continuum. Folding can take place at regular intervals or arbitrary ones. Folding also reverses part of the graphic image and allows visual alignments and coincidences to appear.

10. *Parallax* involves the use of perspectival systems to indicate more than one viewpoint in a model or display. It can be used within representational as well as modeling environments to break the singularity of

presentation in graphical environments. The degree of parallax, or differential between positions or points of view, can be generated systematically or arbitrarily. Points of view can be unassigned, or not attributed, and used as interpretative tools according to the shifts of eye line, horizon line, and other structural features.

11. *Split* is a division within a document, scene, argument, or image in order to show contradictions within the evidence or rhetorical direction of an argument. If parallax registers multiple viewpoints, split can be used within a single author's arguments to refract its multiple aspects.

Other spatial moves can be used to model interpretative actions, such as stack, slip, shift, etc. In other words, any graphical action that can add a dimension to the visual field can be given meaning (semanticized). Understanding the links between these moves and their semantic value will require research and study, user testing, and experimental project modeling. Dimensions are used in combination with activators and inflectors.

The idea of graphical enunciation

Though innovative in many regards, particularly in some of the dimensions and activators/inflectors, the system should be designed so that it can become intuitive, particularly if users understand that a limited subset of features—layers, annotations, semantic inflectors—can serve as a starting point. The more conceptual features—such as fold or parallax—will probably need time to develop habits of use and legibility. The concept modeling environment is driven by the conviction that interpretative work can be supported by graphical means, and that these means can supplement existing visualization conventions of representation and display. By providing a primary and direct method of doing interpretative work, nonrepresentational modeling augments existing mechanistic visualizations with approaches that are much closer to traditional interpretative methods in the humanities.

The development of a means of systematically inscribing enunciation in visualizations is essential to advancing the interpretative agenda of the humanities. The absence of conventions to mark the "speaker" of a visualization, whether it is a display or a model, hampers the development of an approach that acknowledges the discursive modality of visualization. The inclusion of point of view inscribes data within an articulate (enunciative) system so that the display is expressed from a specific historical and authorial position. This allows for contrast, parallax, and multiple views into the same data. User-specific (enunciative) markers allow historical, cultural, or other positions to be registered in the display. Author attribution, the explicit use of positionality as a locator for the speaking (enunciating) subject, and the analysis of the spoken (enunciated) subject, all need to be marked graphically in features that expose the structuring activity of graphical expressions.

Thinking about data visualization as an enunciative system is essential, but it is not a familiar convention of current visualization systems. In addition to mining the conventions of pictorial image production, the project draws on narrative theory, linguistic theory, film theory, critical race and gender studies, and postcolonial analyses where concepts of subject formation (and enunciation) have been developed. Not all of these are text based, and in fact the pictorial conventions provide well-developed foundations for graphical enunciation systems. Developing a set of graphical markers for such a system is part of the research ahead for reasons that should become even more clear in the next chapter.

4   Interface and Enunciation, or, Who Is Speaking?

If a humanistic interface were designed, what would be its distinguishing features and how would it differ from a standard interface used for news, information, entertainment, education, or any other domains? The idea that an interface might expose the workings of its structuring capacity (its enunciative stance)—the creation of subject positions within its modes of address—is far from the realm of common understanding and use. To put this in other terms, if we imagine that an interface is *speaking* to us, then who is it that speaks? How does their mode of address—formal, informal, familiar, seductive, aggressive and so on—make itself evident and create expectations about our relation to that "speaker"?

Critical discussion of interface is rich and draws on theories of media, user experience, and information design with intellectual tools from multiple fields.[1] Reviewing these will provide a way to lay the groundwork for taking the discussion of interface into a dialogue with linguistic analyses of subject formation and seeing how these might apply to graphic dimensions of the interface environment. In addition, we will take up questions about how a humanistic approach might inscribe ambiguity, complexity, and multiplicity of viewpoints to serve interpretative agendas as well as informational ones. This approach links an analysis of interface to the mechanisms by which ideological work is performed in current conventions while also suggesting new features and critical capacities.

One of the themes of this discussion will be the way delusions of agency are produced in interface. In particular, the consumer model of an omniscient individual, gratified through designs meant to emphasize *im*mediacy and transparency, might be considered at odds with the ethical and political identification of subject positionality. If we are subjects of our own experience, situated within the conditions of our own identity formation and its illusions, then what is an interface *doing* as it enables the mediating process of transaction and exchange? How are we inscribed within it? In spite of all of the concerns about privacy and information gathering, individual users are barely aware of the way their identities as information personae are being constructed. Such instrumental concerns are still far from the questions of how the graphical structure of interface design works to produce subject positions. But as surely as we understand something about our identities through a relation to the scale of doorways, chairs, rooms, and other objects in physical space, we also experience our identity through the structure and scale of virtual and digital spaces. An interface structures what we may say/see/hear and how we may navigate in ways that subtly and not so subtly construct our sense of possibility. How might we formulate an approach to interface design that begins with these theoretical issues?

The motivation here is simple. The authoring and reading environments for interpretative scholarly work are only just beginning to be designed in such a way that the linear, finite conventions of print media can be changed for the constellationary, distributed, multifaceted modes of digital media. Early visions of a "multiverse" within digital environments have not materialized. As this process develops, a challenge for humanists is to reflect on and articulate the theory of interface that underlies the design of our working environments.[2] Using principles of interface theory, frame analysis, and enunciation, we can consider ways to structure and reveal interpretation in a screen environment. The terms of formal analysis (basic features such as proximity, overlap, hierarchy, dependency, navigation, and sequence) are insufficient without a model of the subject and agency. How does enunciation become structured in the graphical interface and communicate a model of the subject in the process?

A crucial distinction will weave through this discussion. I am sketching a contrast between an engineering approach that seeks to "optimize" user experience when that user is conceived of as a consumer, and the alternative approach. By contrast, this tries to call attention to the structuring features of power, ideology, and subject positionality at the basic level of the framing operation of interface design. Interface situates us in a mediating relation to information, communication, and experience. It also provides a platform for interpretative work in knowledge production. These positions will be identified in the contrast of the use of the term "user" for the first and "subject" for the second. Interface, in this approach, is construed as the site of "understanding" central to critical hermeneutics.

Though we still have some way to go before arriving at the graphical conventions that will serve our purpose, the intellectual basis for this design is within reach. So is the formulation of this theoretical foundation: the constructivist subject of the digital platform emerges in a codependent relation with its structuring features. This is the "subject of interface" when interface is conceived as a dynamic space of relations, rather than as a "thing."[3]

Defining interface

What is an interface? The term can be broadly construed to include almost any environment for mediating tasks, activities, communication, or user experience from bathroom taps to airplane cockpit dashboards and gaming controls. At a TED conference in 2006, Jefferson Han demonstrated an NUI—a "natural user interface"—meant to suggest a future mode of engagement through multitouch, haptic, features.[4] The idea was that such an interface might be projected onto a surface or object, activated by direct manipulation, and work without being limited to a single device or screen. The demonstration showed him pinching and shaping a glutinous mass with his fingers as if the movements of his hands were being transferred directly through a screen—as if there were no screen. Another demonstration by a Korean company, Celluon, showed the projection of a keypad onto someone's hand and the direct activation of it through touch.[5] You become your own phone and the line between mediation and *immediation* disappears. More recent versions of NUI have been developed on "bump tables" that allow for handling of surrogates for objects and documents by extending the already familiar features of swipe and pinch. The NUI starts to come close to an early vision of a pioneer designer, Don Norman. He "envisaged an ideal 'invisible computer' with an interface that would allow human-computer communication to be as trauma-free as possible. By constantly adjusting to the specific operating logic of humans and machines, the space between user and technology would become an empty space."[6]

Whatever way these conventions evolve, the NUI is anything but a "natural" object. That very designation feels quite Orwellian, given the complicated infrastructure involved—and involving its users. The identity of an interface remains a question—*what is it?*—even while descriptions borrow from paradigms that construe it as a space, a thing, and a process. Under-standing the arguments for these characterizations lays the foundation for a critical humanities approach that also addresses the illusions of agency built into the standard GUI. These illusions are not just the expression of engineering efficiency, but are rooted in other motivations for transpar-ency. The cultural imprint of this engineering approach to user experience remains one of the contributing factors in the historical development of expectations for interface. The assumptions about who users are and what the goals of the interface should be have not been fully questioned by humanistic principles. Even progressive organizations and projects—I am thinking for instance of Mukurtu, a platform designed to serve indigenous communities in web-based presentations of archival materials—limit their modifications to questions of intellectual permission and control of con-tent, not to the ideological effects of structuring features.[7]

How did we get here?[8] Interface is a relatively recent concept, and use of the term appears to rise only in the 1960s, with scant mention in the decades prior.[9] Pioneering work involved virtual flight simulators, head gear, foot pedals, and other devices that disciplined the body. Ivan Sutherland and Douglas Engelbart, two of the pioneering figures in the field as it began to emerge in the 1960s, realized that the challenge of human computer inter-action was to get human beings to have real-time relationships with these "machines."[10] Engelbart designed the mouse, and worked with engineers to calibrate the reaction time of objects on the screen and the movements of the hand.[11] The design of interface ever since has involved the challenges of providing satisfying responses, sounds, and rewards to correlate interaction and results.[12] Interface design is largely driven by values of efficiency and ease of use, and is focused on feedback loops that minimize frustration and maximize satisfaction with mouse clicks, joy sticks, and bells or whistles.

The visual aspect of graphical interface connects it to the embodied con-dition of users, a fact that was evident even in the early days when Engel-bart and Sutherland were struggling to figure out how to use hands, feet, body movements, and orientation to the screen as part of the basic com-putational apparatus. The tactical, haptic, and acoustic aspects of interface have only intensified in the intervening years, though the graphical fea-tures that organize interface remain essential to our use of digital environ-ments. Most recently, the American Psychological Association has become concerned with the addictive character of game and other online interfaces and their effect on children.[13] We have been all too successful, it seems, in creating environments that satisfy their "users" in immediate and absorbing modes. Do these serve the humanities and its critical agenda? Not in their present form.

Interface theory

If we think of interface as a thing, an entity, a fixed or determined structure that supports certain activities, it tends to reify in the same way as a book, which we so often take for a static physical object.[14] But we know that a codex book is a dynamic object, not a static thing. It works through a structured set of codes that supports or provokes an interpretation that is itself performative. Interface theory also has to take into account the user/viewer, as a situated and embodied subject, and the affordances of a graphical environment that mediates intellectual and cognitive activities. Roger Chartier referenced this concept of embodiment as the "engagement of body, inscription in space, relation to oneself and others."[15] In "The Places of Books in the Age of Electronic Reproduction," Geoff Nunberg cited earlier work by Chartier and made his observations about embodiment relevant to the then still very new questions of electronic surrogates and displays.[16] Recognizing embodiment, though important, only gives us a place from which to begin thinking about cognitive processing. It does not supply a basis for a critical theory of interface. Almost thirty years ago, Brenda Laurel defined interface as a surface where the necessary contact between interactors and tasks allowed functions to be performed.[17] She noted, as well, that these were sites of power and control, infusing her theoretical insight with a critical edge lacking from the engineering sensibility of most of the human-computer interaction (HCI) community. And in 1989, Norman Long, a sociologist responsible for social interface theory, described it as "a critical point of interaction between life worlds."

More recently, Alex Galloway (2012) has suggested that we consider a process-oriented approach. Critic Patrick Jagoda summarized Galloway's work this way:

> Considering the interface as an autonomous zone of aesthetic activity, guided by its own logic and its own ends, [Galloway describes] *the interface effect*. Rather than praising user-friendly interfaces that work well, or castigating those that work poorly, [his work] considers the unworkable nature of all interfaces, from

windows and doors to screens and keyboards. Considered allegorically, such thresholds do not so much tell the story of their own operations but beckon outward into the realm of social and political life, and in so doing ask a question to which the political interpretation of interfaces is the only coherent answer.[18] An interface, Galloway argues, is not a stable object; it is a multiplicity of processes. In other words, an interface is not merely a laptop LCD or a television screen. It is not the Windows 8 operating system or Mac OS X. It is not a hyper-mediated heads-up display of the contemporary videogame with its myriad forms of information (health levels, map position, speed, time, messaging options, and so on). Galloway mentions many such objects in *The Interface Effect,* but does not dwell on them. In the first place, he observes, media studies scholars have too often privileged screens and displays. This disproportionate focus on visual interfaces ignores other critical objects, such as "non-optical interfaces" (keyboard, mouse, controller, sensor); data in memory and data on disk; executable algorithms; networking technologies and protocols; and the list continues.[19]

From this perspective, we can see that interface is a dynamic space, a zone in which reading takes place and in which ideological operations are concealed. We do not look through it (in spite of the overwhelming force of the "windows" metaphor) or past it. The desktop metaphor at least suggests a space of activity in which icons stand for objects with behaviors we enact, but it is also a screen that provides an illusion. The desktop makes us believe we are seeing the workings of the system, but the metaphors are not the system structure or architecture. The surface of the screen is not a portal for access to something that lies beyond or behind this display, it is an illusion. Intellectual content and activities do not exist independent of these embodied representations, however, and the metaphors also matter. Interface, like any other component of computational systems, is an artifact of complex processes and protocols, a zone in which our behaviors and actions take place. To reiterate, interface functions as the site of *understanding,* setting conditions for the ongoing negotiation of subject and experience.

If we attempt to separate what we think of as "content" from the wireframes and display techniques, then we are naive. We cannot read content independently on a screen any more than we do when we read the newspaper. If we strip away the graphical codes of a printed text—put its letters and words into a simple sequence, remove paragraphing, hierarchies, word spacing—then we see how dependent we are on these format elements as an integral part of meaning production. They are part of the content, signaling distinctions and embodying codes of reading. These features perform a quasi-semantic function, not merely a formal or syntactic one. The specific qualities of the encoding that distinguish the many modalities of the electronic environment intensify the process of jumping from one frame and media type to another. Distinctions among modes (by this I mean the ways we sort out whether something is an advertisement, editorial copy, or something else, as well as distinguish audio from video, etc.) are largely signaled by various graphical and formal codes that are readily recognized through their conventions.

Within the mainstream HCI community, however, interface is approached through an engineering sensibility driven by mechanistic pragmatism. At its best, this approach is highly and usefully analytic, as exemplified in a much-cited graphic by Jesse James Garrett.[20] This presented the elements of user experience as a duality between the web as an information space and as a task-supporting environment. His observation that the difference between these conceptions leads to confusion in design has fundamental implications for interface. Garrett's insight gets to the basic tension between the display of a rational organization of content (images, documents, "about" pages) and the need to balance this with an intuitive guide to the behaviors users engage in to use that content (search, filter, browse, etc.). An interface does not just show what the content is in a site, but instead, organizes a way for it to be used. Interface is the space between these two: neither the transparent and self-evident map of content elements and their relations, nor simply a set of tasks. Garrett's scheme of organization provides an essential insight, but a full theory of interface goes beyond the design of information structures and tasks into the realization that these are only the armature, not the essence of that space of provocation in which the performative event takes place.

An interface, therefore, is not so much a "between" space as it is the mediating environment that makes the experience, a "critical zone that constitutes a user experience." I don't access "data" through a web page; I access a web page that is structured so I can perform certain kinds of queries or searches. We know that the structure of an interface *is* information, not merely a means of access to it. I may use the search and the query boxes, click on links or drop-down menus as if they are merely a means to an end. But each structure has implications. Sliders, for instance, create a smooth continuum of values. A dialogue box requiring numbers imposes a model of discrete values. When we are looking for dates for travel, it will make an enormous difference whether we are able to state our request in discrete or continuous terms. Interface designers are constantly determining which strategies to use appropriately and effectively. The design of interface is permeated by analytic techniques, and excellent guides to effective interface take into account short- and long-term memory, cognitive capacities, and other fundamentals produced according to the rigors of empirical experiment and user trials.

Interface design also draws on cultural analysis and recognition of differences in symbolic, semantic, and behavioral attitudes. Early work in this realm was done by Aaron Marcus and Associates that studies front pages and their relation to various cultural factors.[21] Building on work by sociologist Geert Hofstede, they looked at the ways cultural value systems are expressed in web design and need to be taken into account. Hofstede's categories are open to contestation, but they provided a way to look at design features. Different cultural groups have different degrees of tolerance for ambiguity and uncertainty, Hofstede's research asserted, they give greater value to individualism or show a preference for collectivism, or register different degrees of dissatisfaction with inequalities in power relations. These features find expression in the graphic organization of information. Interactions with interface would, presumably, exhibit some similar features. Marcus's group did not look at navigation through information structures or at the web architecture to see if these contrasts had implications for use. They remained focused on iconography and layout. But navigation paths, search and query results, browse features—in brief, every aspect of the web content management and display embodies values that are inherent to the reading process, even if they are largely ignored or treated as transparent, neutral, or invisible. Concern with accessibility across a range of human conditions has also prompted self-conscious reflection on the often taken-for-granted aspects of interface design including color, sound, text-to-voice, and remediating features used to compensate for physical impairments. These considerations are important for their work in undoing assumptions of universal users, and the emphasis on embodied and specific individuals challenges conventions and norms in productive ways.

Nonetheless, for the most part HCI designers work to produce effective environments, ones in which satisfactions are balanced with frustrations and efficiency can be maximized. Their focus is on the literal structure of the design, the placement of buttons, amount of time it takes to perform a task, how we move through screens, and so on. Take a typical example of how-to instructions on design: "The Theory Behind Visual Interface

Design" (2002), by Mauro Marinilli, lays out a comprehensive mechanistic approach to the stages of action involved, from "forming an intention" and "specifying an action" to "evaluating the outcome."[22] Marinilli's approach reflects on the design process in relation to a concept of "user experience" that attempts to map structure and effect directly. This is akin to doing close readings of a text's formal features as if that explained the full meaning of a text. This approach to textual analysis, associated with New Criticism in the early decades of the twentieth century, is no longer considered adequate to address the broader issues of cultural difference and individual acts of meaning production.

How do we move beyond a theory of interface based on the "user experience" approach, with its mechanistic emphasis on efficient accomplishment of tasks and concept of the user as an autonomous agent controlled by feedback loops, to one grounded in a theory of the subject? Challenges to the user-based conception arise when we imagine interface engaging with an activity that may or may not be goal oriented, highly structured, and/or driven by an immediate outcome. Consider the diversionary experience of wandering, browsing, meandering, or prolonging engagement for the purpose of pleasure, keeping boredom at bay, indulging in idle distraction, or time squandering. Reading, researching, following trails of association are not necessarily aimed toward efficiency or completion. Artists working online, such as writers represented in the Electronic Literature Organization archive, for instance, are not necessarily interested in *efficient* design, but rather *experiential* design.[23] The same can be said of porn sites. Some radical interventions, such as those created by the art collective jodi.org (created in 1994 by Joan Heemskirk and Dirk Paesmans to subvert expectations of web-based experience through software modifications), directly challenged the engineering paradigm. These aesthetic dimensions, like scholarly pursuits, make interface a space of being and dwelling, not a realm of control panels and instruments only existing to be put at the service of something else. I bring up these contrasting communities because they shatter the illusion of interface as a thing, immediately making it clear that a theory of interface can't be constructed around expectations of performance or tasks or even behaviors.

As long as we think of interface as an environment for performing tasks, we do not approach the question of what other kind of work experience might be constructed. Ben Shneiderman's mantra, "Overview first, zoom and filter, details on demand," assumes that one is working in a very restricted, highly structured, and discrete environment toward a very limited outcome.[24] Shneiderman's user is a *consumer*, not a *producer*, and certainly not a *subject*. We can track Shneiderman's attitude from the innovative beginnings of a robust industry of interface design focused on scenarios that chunk tasks and behaviors into carefully segmented decision trees designed to abstract their use from any whiff of ambiguity. "Analysis," "prototype," "user feedback," and "design" are locked into endlessly iterative cycles of "task specification" and "deliverables."[25] This language is not specific to interface theory, but derived from the more general platform of principles in the software engineering industry. By contrast, we need to consider a "subject" whose engagement with interface in a digital world could be modeled on the insights gained in the critical study of the subject in literary, media, and visual studies. A theory of interface for the humanities might well return to the work of Kaja Silverman, Paul Smith, Stephen Heath, Laura Mulvey, Margaret Morse, and the many other writer-theorists whose synthesis of structuralist and poststructuralist approaches created an understanding of enunciating and enunciated subjects (the speakers and the spoken in film and textual studies).[26] A humanities theory of interface begins with the theory of the subject formulated by psychoanalytic, linguistic, and textual studies and their connection to theories of enunciation in which we learned that "we are spoken by the text" as much as we "speak" it.

But before we turn our attention to the design challenges involved in creating humanistic interfaces, a brief detour through a discussion of frames and framing devices is in order.

Frame analysis

Frame analysis, as outlined in the work of Erving Goffman, is particularly relevant to the study of a web environment.[27] We are constantly confronted with the need to figure out what domain or type of information or communication is being presented and what tasks, behaviors, or possibilities for engagement it offers. By itself, a typology of graphical elements does not account for the ways in which format features provoke meaning production in a reader or viewer. The cognitive processing that occurs in the relation between such cues and a viewer is not mechanistic. Graphical features organize a field of visual information, but the activity of reading follows the same probabilistic tendencies discussed in an earlier section.

These tendencies depend on embodied and situated knowledge, cultural conditions, and training, the whole gamut of individually inflected and socially conditioned skills and attitudes. Frame analysis is a schematic outline that formalizes certain basic principles of ways we process information into cognitive value—or go from stimulus to cognition. Filling in the details of ideological and hegemonic cues, or reading specific artifacts as a production of an encounter—the production of text (reading) and production of a subject of the text (reader)—is a process that depends on specific cases. But the generalized scheme of frame analysis puts in place a crucial piece of our model of interface. This is the recognition that any piece of perceived information has to be processed through a set of analytic frames that are grounded in cognitive experience in advance of being read as meaningful. We have to know where we are in the perceptual-cognitive loops—what scale the information is and what domain it belongs to—before we can make any sense of it at all.

In a networked environment, such as an iPhone, for instance, the literal frames of buttons and icons form one set of organizing features. They chunk, isolate, segment, and distinguish one activity or application from another, establishing the very basis of expectation for a user. Engagement in the interface is an ongoing process of codependent involvement. But reference "frames" are not exactly the same as these conspicuous graphical instances. Once we move away from the initial menu of options and into specific applications or digital environments, a user is plunged into the complex world of interlocking frames—commerce, entertainment, information, work, communication, etc. Their distinction within the screen space and interface depends on other conventions. For scholarly work, the ultimate focus of my inquiry, the relation among frames has some analogies to what are traditionally considered text and paratext. In a digital environment, those relations are loosened from their condition of fixity and can be reorganized and rearranged according to shifting hierarchies of authority and priority. A footnote to one text becomes the link to a text which can become the primary text in the next window or frame, and so forth.

The basic tenets of frame analysis depend on a vocabulary for describing relations (rather than entities). Frames by definition depend on their place within a cognitive process of decision making that sorts information along semantic and syntactic axes—reading the metaphoric value of images and icons as well as their connection to larger wholes of which they are a part.

In traditional frame theory, certain behaviors are attributed to relations between frames. A frame can extend, intensify, connect, embed, juxtapose, or otherwise modify another frame and perception. The terminology is spatial and dynamic. It describes cognitive processes, not simply actions of an autonomous user, but codependent relations of user and system. In invoking frame analysis as part of the diagrammatic model of interpretation, we have moved from a traditional discussion of graphical formats as elements of a *mise en page* to a sense that we are involved with a *mise en scène* or *système*. The shift here is from emphasis on features of layout, or graphic structure, to one that takes the conditions of production into account at a system level.

This puts us on the threshold of interface and a theory of constructivist processes that constitute the interface as a site of such cognitive relations probabilistically, not mechanistically. User behavior cannot be controlled or fully directed because users are subjects, individual and complex. Only by taking into full account the constructivist (created) process of codependence that is implicit in frame analysis can we move from a simple description of graphic features—as if they automatically produce certain effects—to a realization that the graphical organization only provides the basic structure of provocations for reading. The conditions constrain and order the possibilities of meaning production, but do not produce any effect automatically. In fact, as already suggested, the very term "user" needs to be jettisoned—since it implies an autonomy and agency independent of the circumstances of cognition—in favor of the "subject" familiar from critical theory. Interface theory has to proceed from the recognition that it is an extension of the theory of the subject, and then modify the engineering approach to interface that is so central to HCI practitioners.

Constructed subjects

Work by Donald Hoffman on perception as interface extended the constructivist approach to human cognition. In Hoffman's analysis of experience he posited interface as the very site of construction. His "interface theory of perception" outlines perception as a constitutive act.[28] Countering the traditional idea that a species' perceptual capacity is its ability to "address the true properties of the world, classify its structure, and evolve our senses to this end," he suggests that perception is a "species-specific user interface that guides behavior." Like the Chilean biologists Francisco Varela and Humberto

Maturana, Hoffman demonstrates that no experience exists a priori. The world and its reading come into being in a codependent relation of affordances. The new affordances of web-based reading are not distinct from this. They are not another order of thing, a representation already made and structured. They present a set of possibilities we encounter and from which we constitute the tissue of experience. The constitutive act, however, in this new environment, puts our bodies—eyes, ears, hands, heads—and our sensory apparatus into relation with rapidly changing multimedial modes. The integration of these into a comprehensible experience seems to have emerged intuitively, since the frames within frames of the web interface provide sufficient cues to signal the necessary shifts of reading modes.[29]

Attending to the mechanics and logistics of interface continues to reify these elements in ways that do not fulfill the outlines of Hoffman's vision. A more useful approach comes from theories of enunciation that consider all transactions as exchanges that structure subject positions. An interface is a space in which a *subject*, not a *user*, is invoked. Interface is an enunciative, or structuring, system. Texts and speakers are situated within pragmatic circumstances of use, ritual, exchange, and communities of practice.

The enunciative or structuring dimensions of a page identify who speaks to whom and in what orders of language and language modalities. We can see these features at work in most common interface environments, such as Google search results pages, but they are not conspicuously marked. Whose voice speaks in the menu of the Google-verse? The corporate entity of Google of course, but who or what does that mean? The very notion of "authorship" in Google is usefully problematic. Tracking the processes of design in the corporate web would reveal many threads of power, pressure, and contradiction. Porn, we notice, is not among the categories at the highest level of the Google menu, nor are games, gambling, or social networking. The "I" of Google who creates the "you" of the user of the search engine has already interpolated the subject into the structure of the page in such a way that certain desires and interests are subordinated, even stigmatized. Orders of graphic modality and enunciation organize an argument.

The top menu bar frames the Google world under a set of searchable categories. The sidebar, with its chronological order, is a log, a history in which the past disappears below and the present continually refreshes above the long tail of the past. A click on the "All results" timeline for a current theme or topic in the news gives a bar chart tracking an event that recently occurred.

Who identified the data that is displayed? After the fact? When? How does the historical trajectory of what is online match attention offline, and where is the full iceberg of the unrevealed narrative? The visual suggests a slow buildup to a present event, but what happens out of sight is not necessarily of lesser import or scale than the flurry of attention generated by the event. Because Google presumes to provide access to "everything" on the web, its enunciative modalities are markedly different from those of the delimited domains of digital humanities projects. But how do their rhetorics of visualization and presentation express an ideology? In every case, who speaks for whom, where, and how in these graphical and textual expressions? What is not able to be said in their forms and formats? What is excluded, impossible, not present, not able to be articulated given these structures? How can menu categories be altered to contrast one group of users' habits with that of others? How do we get perspective on our own view? Point of view is structured into interface design but never exposed or marked conspicuously. To address these issues adequately, we need to turn to theories of enunciation.

Enunciation

The term enunciation is associated with the work of Émile Benveniste. In 1966 he published an article in which he outlined the basic principles of the theory of enunciation.[30] Drawing as it did on a long tradition of structural analyses, including work by Roman Jakobson and others with roots in Russian linguistics, "La nature des pronoms" (The nature of pronouns) succinctly formulated the generative and codependent structure of subject positions in discourse.[31]

Benveniste's debt to structural linguistics is evident in his analysis of the *exercise* and workings of language. The pronoun "I" always functions, in his analysis, as a link between the system of *langue* and the instantiation of *parole*, linking the two but only as a *locution*. The idea of locution depends upon the situation of utterance or expression being incorporated into the meaning of the text. The "I" has no semantic fixity, no single referential value. It does not stand for or represent any particular concept or entity and has to be understood situationally. (Again, echoes of the critical concept of *understanding* weave through here.) While Ferdinand de Saussure insisted that all linguistic elements are arbitrary and have no intrinsic meaning, the illusion of the representational structure of language persists in daily use.

The term "tree" attaches itself to images, ideas, and things in the world, even though the word "baum" or "arbre" serve equally well. But certain words in language are radically divorced from this illusion, these are pronouns and deictics, terms that can only be given value within the context of their use.

In this article, far-reaching in its influence, Benveniste argues for the specificity of pronouns in their structuring roles. The implications of this analysis for the critical engagement of power in discourse are profound and will be the basis of the discussion of information as enunciation that follows. Benveniste argued that the pronoun "I" identifies the speaking subject, the source of the discourse. But the I implies, and often directly states and uses, the "you" which is the term around which the "spoken" subject coheres. The spoken subject is formulated through address, through direct and indirect implication. Thus, "I is the individual who utters the present instance of discourse containing the linguistic instance *I*."[32] By taking the always implicit and often explicit situation of "address" into account, one has the symmetrical definition for *you*: "the individual spoken to in the present instance of discourse containing the instance *you*."

Put simply, theories of enunciation, and the methods that arise from their formulation, make clear that any communicative expression is an act in which someone speaks to someone for some purpose through some structuring means that organize power relations as positionality. The I/you, we/they, you/others of speech acts are markers of enunciation. They identify the speaking subject and the spoken subject, the enunciated subject positioned by a discourse and constructed, in part, as a projection of it.

In the 1970s and 1980s, the concepts of the *speaking* and *spoken* subject of enunciation were applied to film, visual arts, architecture, and literary works in all genres and forms. But the artifacts and documents of information—charts, graphs, spread sheets, data formats and expressions—were only rarely, if ever, considered within this critical framework. The forms in which data are structured and presented rarely include any indication of speakership. Even when the contents are given author attributions, the structures themselves—spread sheets, repositories, interfaces, and visualizations—are not marked to expose the positionality structured into their communicative work. Visualizations appear as declarative statements, mere presentations of quantitative data.[33]

An Excel spread sheet, a bar chart, and an interactive interface for query in a faceted browser are common forms within information discourse (where the term discourse is a rubric that includes various modes of human expression including but not limited to or defined exclusively by linguistic ones—thus including, for example, filmic, visual, spatial, and physical modes of organized and systematic expression). In each of these the enunciation of subject positions can be systematically elucidated. What is at stake in this analysis is exposing the power structures built into the formal apparatus of the discourse formations that usually go unnoticed and unremarked.

How, then does an information visualization or interface work as an enunciative expression? In what sense can this formulation from Benveniste be used to explore the enunciative workings of interface: "Consciousness of self is only possible if it is experienced by contrast. I use *I* only when I am speaking to someone who will be a *you* in my address. It is this condition of dialogue that is constitutive of *person*, for it implies that reciprocally *I* becomes *you* in the address of the one who in his turn designates himself as *I*.[34]

Who "speaks" an information expression or interface? Who is the spoken subject? Where in the various frames of content does an enunciative function get marked? Authorship goes unmarked on many levels in an Excel spread sheet or browser window—or even a research portal designed for scholarship or within a cultural institution. The structuring of its fields, naming of the rows and columns, semantic designations, and decisions about metric standards for quantitative entities are all authored activities. Beyond an attribute within the filename, or its visual expression contained in a letterhead, a logo, or its location, the spread sheet does not appear to embody the authorial entity of its making, nor of its intended "other" of reception. The browser window is so ubiquitous that we ignore the identity of its corporate sponsor-designers. We might ask the exact same question of a bar chart used to display the data in the spread sheet (or some subset of it) as we do of an interface or a literary text. Where, in the flat, two-dimensional appearance of a graph, flat to the screen on which it is displayed or isomorphic to the sheet of paper on which it is printed, does the inscription of point of view, subject positionality, appear?

Critical interventions

The feint of "unmarked" discourses is that they appear neutral, objective, unauthored, and appear to "speak themselves," to use the language of 1980s deconstruction. Marked texts, those whose discourse positions are framed explicitly, identify the situation and positions of the speaker, as Gérard Genette clearly demonstrated. But the face of the screen, its very *facingness*, is a dramatic embodiment of subject positionality. The information on the screen and the screen itself position the reader/viewer as the "you" of its expression. The plane of expression, the discursive field, on which a bar chart, graph, or even spreadsheet, is inscribed embodies an implicit mode of address. This is a demonstrative and declarative mode, direct in its address. The absence of marks of authorship as ownership or attribution, that is, the absence of signatures or identifiers within the text, is secondary to the absence of attribution on the structural level. But at the structural level, the relationship of enunciation is not only evident, it is overpowering. The facing screen performs an absorption of the gaze that veers between confrontational and consuming modes of engagement.

The structured relationship of the situation of viewing, to paraphrase a concept from film theory of the 1980s, can thus be analyzed as an aspect of display in a viewer's engagement with a page, screen, or wall display. This is the most fundamental use of spatial relations as a way to create subject positions. The screen speaks. The viewer is positioned as the "you" who receives the "I" enunciation of the "speaking" screen. At the level of the artifact, the spread sheet and the bar chart (interface needs its own discussion on account of the many devices it uses to engage and produce a subject) are constructed according to graphical systems that imply an omniscient eye. They have no vanishing points; they appear to be spoken without any historically situated or culturally specific location or agent. This omniscient view is sometimes modified through the use of orthographic and/or axonometric conventions that activate a third dimension along a *z* axis without incorporating perspectival systems that identify the position of a producer and produced subject. The absence of perspective does not obviate the presence of point of view. Point of view is always present. An image, a graphic, or a statement—all are always produced from some place by someone who is positioned in, inscribed in, the structure of the image presentation (even if, as in the caveats above, that "someone" is a distributed, aggregate, collective, corporate, and otherwise non-individual entity). This is an inescapable aspect of human discourse.

The fact of point of view, the inescapable realization that it is present even in the extreme case of the unmarked text or graphic, then pushes the argument in two directions. The first is easy—to address the question of why, then, data expressions, information formats, and their visualizations have so long been unrecognized as enunciative instruments. The answer is simple. The apparent neutrality and objectivity of such graphical expressions aligns them with the ideology of empirical sciences, as if a mere statement of quantitative fact, or of order and arrangement, had no rhetorical value in its own right, but merely served as a communicative vehicle for delivery of information. Discussion of the contrast between delivery/vehicular and constitutive/constructive models of communication has a long history in critical media studies and does not need review here. But the application of its principles to the analysis of information forms and formats is essential and useful.

The second direction is harder. What are the ways point of view and subject positions can be exposed? Can specific modalities of subject formation be shown, marked, within these graphical and formal expressions? Can it be made evident that they *are* enunciative expressions? If we recognize information expressions as enunciative systems, and are able to point to and demonstrate that they are spoken and speak a subject position, then does that do enough to push the arguments about information and/ as enunciation? Or do we need to push further into the ways specific subject formations occupy particular locations within the power structures and regimes that subject us to their operation and controls? If we are produced as subjects of these systems, then they must be specific in each instance, and, as historical subjects, we occupy the relation to them differently, individually, even though we are spoken within their enunciative modes. Even as efforts at content moderation are being put into place, along with debates about their efficacy and ethics, the structuring apparatus of interface is being ignored.

At stake is the need to replace the unmarked activity of information expressions with a systematic critical attention to the "by whom for whom and for what purpose" interrogation to which they should, must, always be subjected. This is a simple hermeneutic move, a shift from reception of these expressions as merely declarative communications to a critical reading of them as rhetorical arguments shaped by the graphical features of their conventions. The size, shape, proximity, order, color, tone, and texture of graphical features are all features in the enunciative graphical system. As sure as tone of voice, inflection, volume, direction, and other aspects of what the twentieth-century semiotician Charles Morris identified as the pragmatics of signs, are at work in linguistic enunciations, so the features of graphical systems are made use of in visual ones (including the apparently minimal graphicality of a spread sheet). All of these features help to position the speaking subject of the discourse within a historical, cultural, temporal, ideological location, and thus to demonstrate the assumptions built into the argument shape that speaks us as its subjects.

If we extend this discussion to the analysis of interactive interfaces, we begin to see the ways in which an even more explicit system of devices for engaging and producing a speaking and spoken subject are at work. The design of interactive interface, the search/query/filter repeat mode of engaging with data, provides an illusion that the structure provides the user with control. The settings in the query can be changed, customized, the search narrowed, the results of the query displayed for a specific purpose. The standard search engine, an n-gram viewer, a repository with a search engine and graphical display—any of these (I'm thinking of Google, Old Bailey, or the Voyant platform for text analysis) provides a set of features that can be clicked, slid, set, have values entered, or selections made. Each of these actions is embodied in an I/you dynamic. The speaker of the system, the "I" embodied in the design, allows "you" certain choices (and of course, by design and by the mere fact of the limits of design, not others). The user is positioned within a disciplining structure that has, in some instances, the appearance of an open-ended system. For a moment, the spoken subject can entertain the illusion of being the speaking subject. I might imagine that I am the author of a Google n-gram query, but of course, on examination, I realize that is not the case. The viewer, its algorithms, and the interface through which I actively engage are all positioning me as a spoken subject of these combined features. The illusion of control, even of authorship, is produced by the interaction with the features, the dials, input boxes, and so forth that all speak me as they speak to me. You, they each say, are the one who does this and not that, or that and then this, and so on. The subject position is enacted and produced.

An urgency and importance thus attaches to the need to address the specific workings of information as an enunciative system. Information—in its forms of expression, formats, and discourses—needs to be seen within a critical frame that strips its apparent declarative neutrality of the stance of self-evident objectivity. Information expressions need to be returned to the realm of discourse to be analyzed for what they are—statements of ideological beliefs that obscure the partial, situated, culturally specific, historically located, and interested nature of knowledge and information as discursive processes.

Simon Penny's useful term "inter-passivity" describes this kind of menu-driven engagement and the illusions it sustains. The designer creates environments that, increasingly, provide a sense of omnipotence. But we are always most deluded when most convinced of our capacity for agency. And the current vogue in click-and-participate forms of activism, social media, and platforms for the rapid registration of opinion are all means of interpellation, rather than fully empowering instruments. When we engage in an illusion, we allow our beliefs to be aligned with something outside of ourselves that we imagine or wish to be true. When we participate in a delusion we internalize that belief and situate ourselves within it. The design of environments of participatory activism raises questions about the model of agency on which it depends—and in what kind of communicative exchange the user-subject of these screens is engaged.[35]

5 The Projects in Modeling Interpretation, or, Can We Make Arguments Visually?

As noted throughout this book, scholars working in digital humanities have engaged graphical modes of data display in the service of research across a variety of fields that deal with the cultural record. Historians, literary critics, geographers, and art historians as well as scholars in the social sciences make use of charts, graphs, maps, diagrams, and other graphical expressions in ways that would have been difficult before the advent of personal computers.[1] Standard programs and platforms enable production of "information graphics" with little or no programming or technical skill.[2] Most, if not all, of the graphical forms built into these programs are borrowed from the natural sciences, or the branches of social sciences for whom empirical models of knowledge production are the norm. In these approaches, graphs use standard metrics, point of view systems are assumed to be ubiquitous or neutral, data lifecycles and sources are not indicated, and other conventions in which display suffices for presentation of information are considered sufficient.

The projects developed here are deliberately designed to explore alternatives to these conventions and their underlying assumptions. They are all based on the conviction that approaches to knowledge that are grounded in user-centered interpretation are fundamentally distinct from those grounded in user-independent approaches to knowledge. Though reductive or crude generalizations that oppose humanist models of knowledge as interpretative and scientific or statistical ones as empirical and positivist are oversimplifications, the simple fact that digital humanists adopted—rather than designed—tools and platforms has led to an unexamined—and in many ways unfortunate—outcome. The graphical forms of visualization are inadequate to the needs of humanists for whom ambiguity, nuance, inflection, and complexity are essential—as is the recognition of the partial, situated, and historically/culturally specific acts of understanding that constitute interpretation. Not only are the materials of the humanities unable to be represented adequately by the points, dots, bars, and lines of conventional charts, even at levels of abstraction, but more importantly, the processes of doing humanistic work—making and modeling acts of interpretation—are not accommodated by the graphical means designed for empirical and statistical sciences characterized by discrete components and disambiguated features.

These project prototypes are designed to show the ways visualization could engage with interpretation across the major areas of current conventions. They all share certain principles because of their commitment to finding alternatives to the graphical forms that serve natural and statistical sciences. The challenge is to create legible conventions for modeling interpretation, rather than to simply continue modeling knowledge or things known.

General principles

These graphical projects are designed to prototype the user experience of six different interpretative activities, each of which is part of humanistic work with the cultural record. The projects are meant to support this interpretative activity, mainly generated either from texts or from textual records which might contain or reference visual or physical evidence, as well as to provide an environment for apprehending the act of interpretation expressed in models. The concept of modeling interpretation depends on the ability to make visible and evident the structure of a model (also understood as an argument) and to be able to compare it with other models. Thus, both the actual work of interpretation and the metalevel work of modeling interpretation are part of these projects' visions.

The projects touch on areas of intellectual work that are directly engaged by digital humanities tools and platforms. Each of these is distinct in the kinds of data or information on which it depends and in the visual conventions with which it is associated (timelines, maps, networks, charts/graphs, ontologies, and point of view systems).

The projects share a few key features, though each uses them to different degrees and can be customized to generate elements appropriate to the project.

- The use of affective and/or nonstandard metrics;
- The use of graphical devices to structure arguments and create data input directly (a nonrepresentational approach);

- The use of inflection and nuance to demonstrate the non-self-identical, nonrepresentational, character of interpretation;
- The use of point of view systems and lifecycles to expose the constructedness of data;
- The use of interactive, direct input, through graphical means—production, inflection, manipulation, transformation;
- The ability to be used as display formats for already structured data or markup.

All of these projects also share some key principles: First, that visualization is a primary mode of knowledge production that can result in structured data outputs rather than be generated by them. Second, that dimensions of interpretation such as ambiguity, comparison, contradiction, and so on need to be legible in graphic conventions. Third, that metrics need to be generated from interpretations, not used to shape them. Finally, that many of the visual conventions from fine arts and design fields (architecture, for instance) could be integrated into these innovative approaches (as per the features discussed in Nonrepresentational Approaches).

The justification for these shared features and principles is that interpretative work in the humanities, as well as the rhetorical character of humanities documents and texts, requires a system of graphical conventions capable of expressing situated, partial, ambiguous, and subtle approaches to knowledge production that are rooted in historical and cultural positions of the interpreter and the objects under interpretation. This user-centered (subjective) approach to knowledge as understanding is at odds with the user-independent approach that characterizes the empirically and statistically driven approaches of the natural and social sciences. The graphical conventions to realize this interpretative approach have never been adequately prototyped. The visualization and interpretation project is designed to take on the challenge of innovating a set of legible features for doing and presenting interpretative work from a humanistic, hermeneutic, perspective.

The projects

The six projects for visualization and modeling interpretation are: temporal modeling; spatial modeling; network inflection; modeling interpretation/argument in an evidence-rich research field; multiple ontologies; and enunciative interface.

Temporal modeling

Two major challenges arise in using conventional timelines to represent the contents, structure, and relations of items that form the cultural record.[3] The first is that timelines designed for empirical data are unidirectional, homogeneous, and continuous. The model of temporality assumed in these timelines is completely unsuited to modeling the complexities of lived, reported, constituted, and produced temporality in human documents, or across assemblages of them. The first challenge is to produce an environment for modeling temporality as relational, not assumed. This requires multidirectional, discontinuous, and nonhomogeneous graphical components. In such a system, temporality, and not time, is modeled. The second challenge is to create a means of presenting multiple and alternative chronologies, or heterochronologies that describe the different chronologies developed to chart historical understanding in different cultural moments and locations. To address these challenges, the temporal modeling project consists of (a) interpretative timelines and (b) heterochronologies.

a. Interpretative timelines   Temporal modeling was my first project in creating alternatives to standard graphic conventions. It was inspired by a project conceived by John Maeda and J. D. Miller in a collaboration done at MIT around 1999–2000.[4] Titled "Grand Canyon," their project was a framework for arranging digital images on a timeline that stretched back from the surface of the computer screen along converging lines of perspective, toward a vanishing point of the past. The idea was to make more images visible by using the illusion of three dimensions, and thus create displays that vanished into deep time and came forward toward the present. The device was simple, and the project was designed for a mass market. The limitations were the same as those of other standard timelines that assume time as an a priori given that is unidirectional in its flow, homogeneous in its metrics, and continuous. These assumptions are just that, assumptions—within the frames provided by subjective experience, narration, or, in another vein, relativity, time is not a given and is not constituted by uniform metrics or unifying conditions. Humanities scholars cannot use these standard timelines to model the temporal experience of novels, films, aesthetic works, or the materials of news accounts, or temporal relations across documents. A radically different approach was needed.

The assumptions in standard timelines have a long history, and as J. T. Fraser, one of the major scholars in the field, has noted, every concept of time and temporality we make use of, with the exception of that introduced by twentieth-century physics, was known in antiquity.[5] The concept of time's arrow and the apparent irreversibility of temporal sequences is central to the human perception of time in experience, as is continuity of time—we do not have gaps, breaks, or jump-cuts in our sense of time's passing. Standard metrics are reinforced by the use of mechanical and digital clocks, but here the traditional models begin to meet some resistance. The basic experience of any standard unit—a minute, an hour, a day—varies considerably depending on the emotional state and circumstances of perception. Some minutes are much longer than others from a subjective point of view, though from the perspective of empirical science and its time-keeping apparatuses, that variation has no relevance.

Taking these three principles of standard timelines and trying to use them as a basis for showing temporality in, for instance, a novel is almost impossible. Time sequences jump around within a story. The narrative combines segments of anticipation and flashbacks. The story is told in segments, chunks, with gaps and breaks. The time of the telling, the literal narration, is never the same as the time of the told, the story. Finally, no standard metrics govern narrative time, which changes scale constantly, stretching, shrinking, becoming compressed or extended according to the requirements of the story. Narrative is a part of other documents beyond fiction, novels, plays, or films and is part of the ongoing structuring of news accounts, historical work, and references within and across documents that form the official record of political and administrative activities. The question of what constitutes an event, what duration it has, and when it begins or ends—these are questions crucial to legal, historical, and diplomatic work as well as to forensic investigation and analysis. Graphical conventions to model this kind of textual and experiential temporality did not exist.

The design process started in early 2000 and the first part of the work consisted of deciding what the primitive components of the graphical system should be, how they should behave, and how a user would interact with them.[6] This tripartite approach underpins all of the projects here, since from a structural point of view, each has to have a set of components, each of which has certain behaviors, and these have to be manipulated by a user in a way that becomes sufficiently intuitive that it can be used as an authoring environment as well as producing graphics that can be read. The interpretive timeline is meant to serve the work of narratologists or other scholars working with situated and experiential temporality in single texts or across a body of texts. This includes any work that depends upon discourse analysis to extract temporal references and frameworks from the rhetoric of statements about when, for how long, and at what rate something will occur.

Interpretative timelines depend more heavily on relational features of texts than on date-stamped information, but they do not exclude one or the other. The basic frameworks of tense analysis, temporal vocabularies, and semantic markup are accommodated by the graphical features (before/ after, at the same time as, etc.). But the interpretative timelines also include models of semantic inflection that allow the codependence or linked influence of events, points, or periods to be demonstrated. The interpretative timeline is meant to be able to work at a very small scale of granularity for narrative analysis, but also work with large corpora (collections) of documents to create analysis and display of temporal concepts across interrelated documents (such as ongoing newsfeeds and their analysis).

The first step was the creation of the basic scheme of components. These consisted of events, intervals, and points, each of which could carry *semantic* and/or *syntactic* inflections.[7] The semantic inflections were basically attributes that could be added as graphical features, like glow, tonal value, color, or texture, to add information to an element. The syntactic inflections linked components through properties like anticipation, regret, causality, influence, or connection of any kind. In addition to these basic elements, the design included a "now slider," which was meant to indicate a viewing point in the present. The temporal modeling scheme had only one now-slider, and it could be progressed to show changes in the temporal model. An event might diminish in importance, an interval might stretch, a point might cast a shadow of anticipation ahead of it and so on. In every instance, the concept of temporality, not time, was in place through the insistence that the model always conceive of temporality as time + a factor of some kind (emotional, subjective, rhetorical, and so on). The behaviors of the components included "snap-to" lines, ability to stretch, layer, fade, and so on. The user was given the ability to add tick marks to lines, rescale, add labels, manipulate objects and elements, and so on. Multiple now-sliders linked to points of view within the platform could support multiple narratives simultaneously.[8]

Once the components were determined and designed, an interface was added to manage various models, layers, and other features of the project.

The infrastructure was built to link the graphical components to an extensible markup language (XML) scheme, but contrary to the usual digital humanities methods of creating markup in XML and then generating a display, this project began with the graphical platform as the primary authoring space. The models were generated from a reading, or a historical event, by constructing the image with the components, their labels, changes over time, influence on each other, and so on. These graphical models were stored as XML and could be exported in that file format.

The built-in limitations of the project came from its reliance on a Cartesian grid. We did not have the technical capacity to generate nonstandard metrics. The regularity of that structure made it impossible to introduce the kind of warping transformations that could indicate affective forces at work. The use of stretchy timelines and also blown-up segments produced through a recursive attention to detail were not possible in the system as designed. Nor did the project develop sufficiently to work in dialogue with XML to generate a display from markup in an iterative process of back-and-forth between graphic modeling and textual markup and XML. This would have supplied a feedback loop between the interpretative work, input, structured data values, storage, display, and so on in an iterative mode.

This project was developed to proof-of-concept stage with workable features, layers, export and storage functions, and an administrative infrastructure for managing models.

b. Heterochronologies The idea of heterochronologies, or multiple timelines created from different cultural or historical understandings of time, arises from some of the same impulses as temporal modeling. In particular, it recognizes the need for nonstandard metrics that embody alternative models of the organizing framework of historical events. The project points toward the idea of being able to contrast and compare ontologies exposed in a graphical format, which will be described below. The understanding of the history of human culture has produced a striking array of different chronological schemes. Some of these overlap, coordinate, intersect, or can be correlated, but others cannot. Respect for the cultural otherness of the past, as well as for the distinctions among different models of historical time, requires a graphical system that does not force alternative chronologies into a single frame of reconciliation. A precedent for this is found in manuscript work by the third- to fourth-century scholar Eusebius, one of the most significant chroniclers of antiquity, where multiple chronologies are exhibited in columns across pages that are not organized by a single underlying or unifying grid. Creating an environment for doing this work within a digital environment is crucial as a representation strategy and also an interpretative one.

Each chronological scheme should be able to be produced according to its own component parts, scale, order, and structure and then layered into a relationship with others (depending on the appropriateness of correspondence or lack thereof across the schemes). So a set of metric (measurement) variables, with the organizing features of each chronology (events, duration, points, milestones, breaks, absences, questions, and so on) will be created as an effect of interpretative engagement. Variable chronologies do not have set start or end points and do not have standard annual or chronological markers or structures, but reflect the models on which they are based or created. Chronologies will have whatever extent and granularity the historically or culturally specific project requires. Fictive chronologies can be accommodated in the scheme.

My need for a way to present alternative models of historical chronology arose directly from work on a text by Edmund Fry, the 1799 *Pantographia*, a compendium of samples of all scripts known to the author (also a printer–type founder) at the time. Fry's extensive research resulted in the elaborate production of specimens, but also recorded their sources along with cited excerpts about their historical origins. Thus, a script might be described as having its source "in ancient paintings," or as having been created "after men had lived long enough in a state of society to perceive the insufficiency of inarticulate cries and gestures." The chronologies on which Fry was drawing at the very end of the eighteenth century were biblical, historical (linked to particular events and secular dates), or keyed to the dating system that depended on the four-year cycle of the Olympiads. These were sometimes stated in relative terms (before, after, during another event), sometimes descriptive (e.g., "revealed from Heaven"), and sometimes linked to specific events ("at the invasion of the Spaniards"). These modalities of relative, descriptive, and specific events were linked to the cosmological, biblical, historical, relative, and other timescales. Fry's historical references cannot be unified into a single system. To do so would do a violence to the cultural otherness of the past and disregard the extent to which Fry's understanding produced a complete explanation of the historical past. The dates in his documents could be coded using a scheme that identified them as belonging to one of the several timescales: C = Cosmological (big time scale, no specifics); B = Biblical time scale (5000+ years, reference to events); H = Historical time scale (actual dates); R = Relative time scale (frame unclear); O = Other time scale. For example, dates like "1713," or "93" are historical; "authors pretend that Moses and the prophets used this letter" is biblical; "brought from Heaven by the Angel Raphael, by whom it was communicated to Adam," is cosmological; "when the Christian princes made war against the infidels" is relative, and so on. A markup scheme can certainly be generated for these values, but they cannot be placed in a grid governed by one metric.

No single timescale works for all of these dates, and none of them conform to the historical timescales that emerged after James Hutton, Charles Lyell, and Charles Darwin transformed understanding of the ages of the earth through their discoveries.[9] The worst possible solution to portraying the highly nuanced and complex information in this historical document would be to map all of its specifics onto a single unified timeline constructed in terms of current models. The bias of presentism erases historical otherness. The result would be to flatten the information in Fry's carefully transcribed citations, and to thereby ignore the very rich understanding that can be derived from them. Fry had no problem making use of multiple chronologies, and he saw them all as valid within their own frameworks. Our understanding of that position depends upon being able to respect and present these heterochronologies in their specificity and being able to fully appreciate their sometimes contradictory and sometimes simply different models of historical time. Again, the graphical scheme requires use of multiple metrics within the same graphical space, and some means of showing points of connection across them. The system is meant to be used for creating and displaying chronological schemes from historical and cultural sources so that they can be compared without being reconciled into a single overarching system.

This project remains in the conceptual phase of development with only hand-drawn or screen-sketched versions of the prototypes. Because chronologies are a form of classification and description, this project is similar in the challenges faced by that of comparative ontologies (see below).

Spatial modeling

Spatial and temporal modeling share a conviction that neither "time" nor "space" are a priori givens within human experience and perception, but are always constructed. That construction can be shown, given form, and demonstrated through visualization techniques. The rationale for this project is to create a systematic approach to portraying spatial information from an affective, or subjective, perspective. The founding principle is that space is never given, only experienced, particularly within the representations of narratives, historical events, performance, planning, and experience. The space of an open square is different at all times of day, on different days of the week, and in different seasons. It is also experienced differently by individuals with different profiles and even by the same individual in different roles. The ability to model these experiential features of space requires that we create an input feature that lets the space be inflected, warped, or otherwise inscribed with values. The space should come into being in part through these means, even if it is indicated in advance in a standard mode such as a map, schematic outline, or other format. Imagine a room emerging through the details and information provided by a narrative, rather than created wholesale. Just as temporality is time modified by a factor, so spatiality is modeled with various factors.

We don't perceive space—or landscapes—from outside them, nor, rarely until relative modern times, see them objectified, as from the air. Though cartographic models are much older than air flight, early maps were often records of encounter, particularly with unknown lands. All maps are projections, and in the well-known truism of cartography, they are all attempts to show a three-dimensional form on a two-dimensional surface. This is true no matter what the scale of the map, though maps that take in larger territories are more conspicuously marked by the distortion of chosen viewpoints than those at smaller scale. The creation of projection systems to compensate for distortion always involves a trade-off, and the compromise to sacrifice accuracy of size for precision of position or relation depends on the use to which the map is being put. Orientation, north-south biases, and other matters are also well-known features of cartographic conventions.

But the concern here is not so much with the standard problems of projection, but rather with the ways in which humanistic documents register spatial information, are used to create spatial models, and are embedded within models of space that, traditionally, do not involve any affective or experiential features. How would a digital humanist expose the spatial imaginary of a work? Is James Joyce's *Ulysses* to be mapped onto a plan of Dublin with pins and points as if the city were a container for the experience? The complex codependencies of the narrative and its relation to spaces, references, and places require a different mode of modeling.

The structure of this project needs to include use of preexisting materials, like maps or plans, photographs or other depictions, of a built space as references, but it should also allow for a spatial model to be built directly from references and evidence, creating variable metrics. In the first instance, imagine a map of a battle or struggle and the ways the terrain is perceived by different figures within the event depending on their familiarity with the space, eye lines within the space, pressures of the events and moments or degrees of danger to the individuals. In the second instance, imagine a creation of a geographical terrain according to the narrative of its discovery. This does not need to be an instance of colonial contact to require a constructive approach; it could be a story of travel or discovery in other circumstances, or the production of a geography from partial evidence. What are the spaces depicted in novels? Films? Historical references? These provide only points of information, not complete descriptions. In fact, no full description of a space is ever available and the mediating process of knowledge of a space is like that of a text or aesthetic work.

The components of interpretative modeling of spatiality need to include ways to indicate the process of constructing a space through acquaintance with it, experience of its qualities, the ways evidence contributes to its emergence as a model. A system in which a map is layered so that only parts of it come into view, or high focus, and only when there is evidence to support the interpretation, would be completely different from the current approach to mapping in which a historical or contemporary map is used as a base image, then marked with pins or points that might also be linked to bits of evidence, but without having any effect on the geographical representation. If a space is created only through evidence, then it calls that process of spatial visualization to attention. Similarly, if a geography is morphed through affective experience, then this should change the coordinates and topography of the map. The idea of heterogeographies depends on an input system that allows a user to stretch, pull, increase, or decrease any component of the metric on which the space is drawn.

The components of a spatial modeling system would include a point of view system that could be occupied by one or more participants. This would allow the spatial model to be viewed from within a subject position as well as from outside it. The features for generating an affective metric (or variable metrics) to morph the standard grid would need to carry information about sightline and view window, area of impact and influence, rate of change and degree of stability of any phenomenon in respect to the space, and a legend for identifying types of events or elements, their degree of ambiance, resistant forces, limits, etc. In other words, any and all features of spatial experience would need to be able to be modeled within the environment. The simplest spatial models might be those created from textual evidence, and gaps in the historical record or narrative would be made clear as would the relative weight, authority, and value of the evidence.

Experiential and evidentiary models are a crucial feature of spatial modeling. So is the ability to create a workflow that would take features of text analysis and topic modeling back into an expressive dialogue with coordinates in a standard system and inflect these so that the landscape morphs. In such a system, frequency of reference would register in some way on the map—simplistically, this could be by a change in size, tone, or position. All cartographic features should be able to be manipulated as a "factor" of their experience and the results should be able to be displayed as a warped, ripped, cut, or partial map.

The two modes, experiential and evidentiary, are not mutually exclusive. In each case, the approach to implementation involves some of the same features. Either the spatial references can be used to build up an image or the evidence can be used against a base map or image. The coordinates on any map project would be malleable, able to be pulled and stretched. Point of view systems need to be enabled so that the space or geography can be seen from a particular perspective, and multiple viewpoints should be able to be embedded in any model. A legend that uses semantic and syntactic inflections, similar to those in the temporal modeling environment, allow for the addition of attributes, as well as labels, or interconnected causal connections and relations (something that expands, spreads, smells, contaminates, or has any kind of effect from one region to another would be connected syntactically). All models need to be able to be advanced through time along a slider or other device. Completeness of any model should be an effect, not an a priori given, and the degrees of focus, granularity of detail, and expansion according to either amount of attention or evidence should be fine-grained enough to allow a tight correlation between information or evidence and the rendering of the image. Spatial models include three-dimensional virtual constructions of space (for archaeological, architectural, cultural, narrative, or other sources to be modeled) and maps (of territory, experience, events, and so forth). In all cases, the affective dimensions of modeling are crucial to the shape of the outcome. Spatial experiences are specific to circumstances, and models should be able to express these specific characteristics. A space and an event can be tightly related in terms of their model, and the relation between occurrences and spatial conditions and features should find expression in spatial modeling.

The basic implementation system consists of a staging space (two- or three-dimensional, with metrics able to be specified in standard and affective formats) and/or a reference image (or images—photograph, drawing, record of any kind, and/or map). This basic space and/or image should be able to be manipulated by and through its coordinates and/or projection system. As mentioned above, point of view and perspectival views should be available to insert the model into an individual and/or shared or multiple viewpoints. Sight lines and view cones should be able to be specified. Geographical and dimensional features should be able to be sketched and modeled. Evidence should be embedded and viewable, and elements/features of the cultural record should be linked to their graphical expression. Textual evidence should be able to be data-mined for coordinates (vague or specific). The use of semantic and syntactic attributes should be customizable, though a basic set of inflections should be built into the system. The use of affective metrics—metrics generated from subjective experience, not used simply to register it—is essential to both spatial and temporal modeling. Variable metrics allow for contrast and comparison while also addressing the problems of rational graphical systems of spatial representation.

Network inflection

All features of spatial and temporal models have potential to be embedded in relations. But in a network, the chief structural feature is the *relationships* among components. The data structure for a network has a tripartite structure so that the relationship can be specified. The standard network visualization consists of a node and edge—a point and a line connecting it. Each can carry weight and also other attributes. These can be indicated by color, size, or other graphical features. However, most network diagrams reduce all relationships to the same presentation and make static representations out of dynamic conditions. They are premised on a characterization of entities as discrete entities, or nodes, that are not affected by the conditions of relations in which they are involved. This is a highly mechanistic characterization of nodes (and edges), whether they consist of human beings, institutions, or events. The model does not allow for any influence or effect to transform either the node or the edge relationship as a result of the dynamics between them, change over time, or any other of the many factors that would feature in an actual relationship. Inflected network visualizations would be dynamic, they would include change across time, and they would be based on a notion of the "nodedge" as a combined node-and-edge effect of codependencies that are the very substance of relationships. The rationale for this project is simple, that the current node and edge model is inadequate to show the many dimensions of relationships, their specificity, distinction from each other, and change over time.

The structure of the project allows graphical input to create attributes and values for the basic data structure. The node and edge can be linked or manipulated independently. Either can be increased or decreased, given attributes (inflammation, steady state, diminishment, intensity, etc.). When the node and edge are linked into a *nodedge* system, changes to one or the other, or to linked nodes, would reverberate throughout the system. Systems can be bounded or open, inclusive or selective, and these features can also be managed with a timeline on which such features change across an interval or at a particular critical moment. The conceptual framework is simple—that networks are living, dynamic, and constantly changing, not static structures comprised of parts that are identical to each other. Input into the models should be direct, graphic manipulation linked to a paint box of attributes, each of which is indexed to a data structure and value scale. Data-mining from texts and direct input from tables with static, multiple, or dynamic attributes will also provide input. If-then statements should be included in the table so that relationships can be modeled according to variable factors and transformations. These conditions could be linked to terms in a linguistic sentiment analysis and also to locations in a text or corpus (e.g., extracting information about reports on the Trump-Putin relationship from news feeds and sources over time).

Modeling interpretation in a rich research (discourse) field
This project integrates the argument space and evidence in large and small corpora. The goal is to create a navigable and legible graphical expression

of the argument space or spaces (multiple pathways, arguments, narratives, and discursive interpretations can be generated from any discourse field). This project also makes use of the tools of topic modeling, data mining, and other analytics as part of the evidence that can be generated and linked to the models. The project seeks to create a visual convention that is legible and allows user interaction and engagement, as well to support active two-way display and modification.

Take a complex combination of documents for a research project and imagine all of the ways it can be used, connected, analyzed, and interpreted. Because multiple, even infinite, potential interpretations can be imagined and produced, the system has to be capable of holding many arguments made from the same evidence. One way to describe this phenomenon is that a discourse field is comprised of evidence and a reference field is constituted by the reading that the discourse field supports. An archive of papers from an institution can be read as a political history, as personal sphere of influence, as a record of critical events and issues, and so on through a varied set of individual readings. Each reading is structured by an interpretative model. An analogy would be to imagine the infinite number of constellationary figures that can be drawn among a field of stars. The stars are the evidence, the documents, the objects in the discourse field. The connections among them are the readings that create a reference field—the image/figure is an argument about how the configuration of the stars can be read.

The basic rationale for this project is to find a way to support and expose multiple readings and interpretations in a discourse field and show and compare the relations made by different arguments among the pieces of evidence. The field itself might be constituted differently for the different readings, and links to collections, materials, documents, and evidence would be differently constructed in each instance. Again, imagine, for example, a repository that documents an ancient site, its bibliographical history, its excavation, and the dispersal of artifacts, drawings, records, reexcavation models, and other materials relevant to the site. Links outward from that repository would vary according to projects, linking it to other sites, excavations, professional practices, theories of archaeology, individual figures, historiographic materials, and so on in an unlimited proliferation of possibilities. Or, it might be part of a study of animal habitat and changing climate conditions. The nonunitary nature of the reference field arises from the fact that no two readings or interpretations are alike. The discourse field, even if it remains bounded and unchanged, would still support a wide range of interpretative projects, like pathways through a complex physical site do. When the evidence is all in view and available, the job of the scholar is to provide a reading, a path, an interpretative introduction to the discursive materials.

The project takes a nonrepresentational (i.e., generative, argument-driven) approach to the graphical display of models of interpretation in relation to a complex discourse field. The analogy of layers of pathways or tours of the materials is appropriate here, though of course the pathways need not be linear, and the display would not be limited to two dimensions.

The project could make use of data analytics as part of the input stream. Topic modeling, text mining, network analysis and other methods of analyzing and presenting interpretative work could be integrated into the display. These would be activated by filters and selective display with point of view and now-sliders to identify author-subject positions. Analytics might pull certain evidence to the fore or expose connections among elements in the discourse field. In this regard, the project would draw on other features built into the discovery tools in nonrepresentational approaches. Allowing direct interaction and manipulation of the relationships and argument structures in the discourse field is a crucial aspect of the project—perhaps the most crucial, as it would allow for an intuitive interface to the problem of modeling interpretation in a rich discourse field. Determining what the interpretative moves are, what the fundamental argument structure should be, and how the use of inflections and attributes can be employed would build on already described feature sets of other projects. The chief difference between those projects and this would be the extent of the discourse field, the open-ended nature of the connections between a bounded field and an extensible one, and the complexity of the model. In order to draw substantive content into the argument space, the graphic display would need to support an authoring function that would be as minimal as annotation and as maximal as full writing practice, with both linear and nonlinear structures.

The project would use the graphical interface as a structuring environment supported by a display/discovery space; the graphical interface allows a user to create an argument that becomes part of the structured data, rather than simply working to display the already structured data in a table, database, or file. It is nonunitary because it supports multiple (infinite, open-ended, nonsingular) interpretations by one or more users. And the argument can be stored as a data structure, independently.

The term reference field (or plane of reference) here is meant to describe the constructed space of interpretation; it is tractable as a text or organization of evidence within an explicit structure; in this sense, iteratively, the reference field becomes evidence itself for the next cycle of analysis but in the first instance, it is created on the basis of the discourse field, which is the collection of evidence—documents, photographs, maps, files, tables, data, any artifact that is being used for research. The work of modeling is the primary, active creation of an argument structure or organization—as opposed to display, which is a secondary expression of an already structured data set.

The problem being addressed by the project is to see if the active creation of argument can be given graphical expression in a way that is legible and intuitive to produce and to navigate. The argument is a stand-alone discourse but it is also dependent on and related to the underlying field of discourse (the evidence on which the argument is based). What kind of visualization would be useful so that the interpretation can be "read" for its argument but also point to the evidence on which it draws? To what extent are latent arguments within evidence able to be exposed through this process as well?

The project has various functional requirements which combine the feature sets of the rest of the projects. These include: the ability to create and express an argument in relation to evidence; links to and anchors in specific pieces of evidence (digital artifacts); the ability to use open linked data and metadata to perform analyses; the ability to use analytics—data mining, text mining—as evidence; the capacity to show the argument as a model in graphical space; the ability to interact with the models and track back to the evidence; the ability to modify the model and work with it responsively; the capacity to maintain and preserve alternate (infinite) readings of the evidence; versioning ability and ability to track back into argument development to show the lifecycle or at least partial history of the development of the argument.

Comparative ontologies

Classification schemes for knowledge organization, known as ontologies, are based on structures that are themselves semantically meaningful and legible. Consider the basic inventory of life forms and their organization by domain, kingdom, phylum, class, order, family, genus, and species.[10] This is a hierarchy organized from general to specific, with each lower level inheriting the properties of the level above. Does it conform to the actual description of living beings and their connections to each other? Yes and no, and the ways in which life forms are classified has transformed considerably under influence of the study of genetic material and markers of species identification, but also, with recognition that the strictly hierarchical system of identification does not account for the way species differentiate. Completely different descriptions of living beings, or parts of the larger domain of living things, organize them according to other criteria—edible or not, taboo, to be feared or not, able to be domesticated, useful for their skin or fur, able to be bred. Cladistics provide a way to assess the percentage of shared characteristics species have and when they might have branched from a common ancestor. The branching structures in these systems do not match the hierarchical trees of the kingdom to species models. The ontologies that describe these categories of description are necessarily different from those that organize them simply as living things with particular morphological characteristics.

In addressing the cultural record within a long historical timeframe and a broad cultural one, a challenge arises with respect to preserving varied ontologies and using their specific features and qualities as legible components of a structured worldview. As organizations of knowledge, they embody and express cultural and historical understandings. The point of preservation is not to force these varied knowledge systems into standardization, but to find a means of exposing their differences and making them legible in a meaningful way. This project addresses the issue of comparative ontologies as a politically charged, culturally necessary problem in the management of the cultural record.

The rationale is simple: in order to avoid hegemonic standardization that represses cultural differences, we need systems that preserve differences and expose the specific features of the organizational structure. The challenge is to conceive of the terms of comparison and to consider how points of contrast and connection can be structured and made legible. Each ontology has to be considered as a whole, and its structuring principles made clear in graphic, schematic terms: hierarchy, table, intersecting areas, multiple discrete parts held in aggregate, etc. In other words, what *is* the whole and how are the part-to-whole components organized? Faceted classification systems, which operate using a nonhierarchical set of features, each with its own vocabulary and criteria, have yet other structural properties.

The variety of ontological ordering systems is much greater than in chronological or temporal ones, though temporal schemes are themselves ontologies—classification schemes for organizing knowledge about a topic or subject area. Consider the problem of creating a comparison between the ways an indigenous community views its artifacts and the ways archaeological researchers or art museums organize these same artifacts. The differences are important, and each organizational scheme tells us something about the circumstances of use and cultural conditions of its development.

Another pressing issue arises from the development of linked open data systems and the need for exchange and correlation among varied legacy systems of knowledge organization. The challenge is not merely that of being able to find and access individual units of information within and across schemes, repositories, and formats. Because the cultural and semantic significance of any item, object, document, or term is dependent upon its place within the larger relational scheme, making a single crosswalk on an item-to-item basis does not adequately resolve the problem of comparison. Consider, instead, an emergent master list comprised of every object/entry or term in every classification system that is part of the comparative scheme. Each access to the term produces a graphic display of the ontologies of which it is a part. The formal structure of these systems could be exposed as a semantic system in which position, meaning, and value are integrated. These could also be filtered for search and display to promote legibility, and various scales of graphical comparison would allow for more and less detailed examination.

Active engagement for organizing documents, objects, or terms into an ontology would be facilitated by an interface that allows an ontology to be created directly through drag, link, place, position actions on the part of a user. This would be particularly useful for archival work and for creating multiple ontologies in a single collection where such issues as provenance, collector's imprint, original order, alternative ordering systems (authors, titles, dates, place of production, chain of reaction/response) might also be usefully developed. Subsections and partial organization schemes within a larger collection could also be facilitated in ways that relate to the argument structure and discourse field described in the previous project. Preserving ontologies from on-the-fly organization would create structured data for preservation, modification, and reuse. The graphic conventions for display would be inconsistent across ontologies, but could generate comparisons based on points of overlap, contrast of scales, and structuring principles, through locating common or shared components and exposing their relative position within a scheme. Learning to read schemes of organization for their graphical arguments would be a side benefit of this project (hierarchical and not, extensible or fixed, faceted or linear, etc.).

Enunciative interfaces

Ideology is always at work in the linguistic components of a site. But the ideology of interface, as described earlier in this book, works in part through the obfuscation of the relation of speaking and spoken subject of enunciation enacted in the mechanisms of display. The apparently neutral graphical structure of a flat screen, a rationally organized display, and conventionally organized space of menus, navigation, and information organization all carry signs of enunciation. All are part of systems that position a spoken subject of graphical enunciation within a power relation to the speaker. How to show these relations? How to intervene in the conventions of graphical user interface to expose its workings? What kind of "reveal codes" feature would show the assumed speaking and spoken dynamics of a site—and also show its connections to other sites, institutions, or authoring and authorizing groups of which it is a part?

This chain of production, very hard to recover, poses a challenge for data analysis as well. Data lifecycles are rarely exposed, and the creation of visualizations usually begins with a data set, rather than with any engagement with how that data came into being. Data are also "spoken"—they are part of systems of enunciation whose workings are concealed.

The interface challenge is different from that of the data set, but both involve creating some mechanism and convention for identifying source, speaker, and life cycle of production. In the case of the interface, a solution might introduce a graphical code that translates apparently neutral organization into zones, and modes of address into a legible system. Just as the markers of "I" and "you" are clearly identifiable in a linguistic utterance, so the terms of address need to be made legible in a graphical expression. The chain of production indicating "who" the speaker is in any instance—where the source of the site resides, who hosts—could be made explicit as well.

For data production, the task is to expose some of the procedures and steps by which data is created, selected, cleaned, and processed. Retracing the statistical processes, showing the data model and what has been eliminated, averaged, reduced, and changed in the course of the lifecycle would put the values of the data into a relative, rather than declarative, mode.

This is one of the points of connection with the interface system and task of exposing the enunciative workings. But data lifecycles and models could be embedded into any graphical display, showing how the original data set and sample were related and what the stages were that led to the data set from which any graphic is generated. This kind of analysis is central to the understanding of electoral processes, consumer patterns, opinion polls, and also the kinds of projects that generate topic models, network analyses, and other outputs from the cultural record. *What percentage of what record, massaged according to what parameters, has led to the data set?* The graphic would have some of the properties of bringing a focused image out of a blurred background, but also aggregating and simplifying into a coarser pixel resolution as things drop out of the data set. Students' horror when they are exposed to the task of data "clean-up" is a clear indication that being able to show data lifecycles is a crucial part of analysis and display.

Data input is structured in tables, style sheets, xml, and html. None of these are conspicuously marked with regard to the way they function as enunciative structures. A graphical input to assign a speaking role from within the interface would add an attribute to the data structure. The display could be filtered according to attributes. Data lifecycles would be shown through a sequence of graphical displays that provide information about the steps through which the data was refined, and the changing models according to which it was structured. Author attribution for data sources, and connections to networked resources that provided the data, counteract the apparent neutrality of data presentation. Standard conventions do not show the lifecycle of data production or the variations and complexities within extant data sets, they reduce numeric anomalies and specificities to generalizations and aggregates. An innovative approach would reveal the multiple dimensions of data, its irregularities, complexities, variations, and the conditions of its production while also inserting point-of-view systems within the data to show that it is an act of enunciation, created from a position within data production, not outside of it.

The projects are structured around a set of identifying features to mark enunciation and distinguish speaking/enunciating and spoken/enunciated positions. Each involves a degree of tracking back into lifecycles of production and authorship. The data lifecycle has a more specific and focused attention on the statistical transformation of data and the models according to which it was created, manipulated, transformed, and presented. The conceptual features of the projects are aligned in their commitment to exposing the unstated authorship concealed in the structure of the graphical display. The apparently neutral declarative statements of interface and data display share an ideological agenda *to simply appear to be what is*. Taking apart the pseudo-transparency by showing the workings and apparatus of the interface and graphical display of data is a crucial act of hermeneutics applied to information displays and systems.

To make these points vivid, we can briefly examine the specifics of cases. Consider the contrast between two digital platforms, both instances of what we term *knowledge design*. The first is an interface designed by Anne Burdick for a project of the Austrian Academy of Sciences. A team of linguists and social scientists worked for twenty-plus years to transcribe, make facsimiles of, and markup (with tagsets), the thirty-seven-year (1899–1936) run of the journal *Die Fackel* (The Torch).[11] Edited by the writer Karl Kraus, the journal charts the peculiarities of language he used to encode criticism in an era of rising fascism. Burdick's interface is a skillful site of mediation between the multiple dimensions of the repository and the functionalities it supports. Her design creates an environment for research that interconnects the facets and features of the various files, formats, and their affordances. The user/viewer of this site is guided into use of the materials—whose content is as substantive and mordant a critique of the role of language in ideology as any ever produced. The viewer's sense of agency is rooted in the legibility and apparent transparency of the interface. In fact, much processing and infrastructure are rendered invisible by Burdick's design, as it is neither necessary nor useful for a researcher to see the file structures or pipeline that push content to the screen. The skills Burdick brought to this project were conceptual and intellectual, as well as graphic. Her knowledge had to encompass an understanding of the many parts, functions, and interconnections of the repository—but not the technical aspects of implementing their design. She had to understand what structured data can do, but not how to structure data herself.[12] More remains to be done to integrate this kind of highly specific approach to location and orientation into visualizations of arguments and evidence.

A striking contrast comes into focus when we compare this site with that of a business systems software "cockpit" and its role in providing a user with a view of a supply chain. In her recent work, digital humanities scholar Miriam Posner has focused on these platforms to show how they produce an illusion of omnipotence—the whole world is shown on the screen and every phase of the process can be searched, found, faceted, displayed, pinpointed in real time.[13] But what the design of these platforms *does* is conceal the costs of production—human, labor, environmental, economic, political, and so on. The agency of the designer is in the service of relations between human beings and intelligent systems whose use of neural networks and other capacities for bootstrapping and self-optimization do not depend on people. Once put in place, they work out of sight and, largely, out of control. Design, in this instance, is not a matter of the creation of communications between individuals, or even companies or corporations, but of the surface display of complex systems with which we do not have the power to intervene.[14]

As we arrive at the end of this critical analysis, we are still left with the dilemma of how to implement design features based on these insights. Can we conceive of models of interface that are genuine instruments for research? That are not merely queries within preset data that search and sort according to an immutable agenda? How can we imagine an interface that allows content modeling, intellectual argument, rhetorical engagement? In such an approach, the formal, graphical materiality of the interface might register the performative dimensions as well as support them. Such approaches would be fundamentally distinct from those in the HCI community. In place of transparency and clarity, they would foreground ambiguity and uncertainty, unresolvable multiplicities in place of singularities and certainties. Sustained interpretative engagement, not efficient completion of tasks, would be the desired outcome. Grounded in principles of interpretation and a theory of subjectivity, such an approach to design has yet to be developed. But it would expose the process of thinking rather than display fixed results of intellectual activity as if they were products.[15]

Rather than conceive of interface as a threshold, a liminal zone, a space between, it should be thought of as a constitutive space of enunciation, not simply a gap between one thing and another thing, a user and an operating system, in a mechanistic sense, but a mediating, transacting, *productive* space. Lori Emerson commented on the ways this approach allows the discursive power of interface to be apprehended. "It also seems to me that an attention to interface—again, made possible through attention to certain works of e-literature—is a crucial tool in our arsenal against a receding present … by which I mean without attention to the ways in which present and past writing interfaces frame what can and cannot be said, the contemporary computing industry will only continue un-checked in its accelerating drive to achieve perfect invisibility through multi-touch, so-called Natural User Interfaces, and ubiquitous computing devices."[16]

The challenge of creating an interface in which the performative character of interpretation can be supported and registered builds on demonstrable principles: multiple points of view, correlatable displays, aggregated data, social mediation, and networking as a feature of scholarly work, and the qualities of games with emerging rule sets.[17] These features, like those discussed in the section dealing with hermeneutics, are elements that fracture the singularity of interface, allow situatedness and constructed positionality to be registered, and force the user to confront their subject identity as an aspect of their experience. Within a humanistic environment, the need to record subject position and authorship and offer multiple perspectives of critical reading and interpretative pathways through the complexities of the cultural record also requires new approaches to interface design.

The interface makes a visual contract between maker and viewer, between the display and the subject positions it constructs. A humanistic interface might, at the very least, unmask those workings and allow subjects to inscribe themselves within its frames, to note, mark, acknowledge, and even respond to the enunciative apparatus, to allow spoken subjects to begin to understand the position they occupy and the illusions of agency this position promotes.

Summary

All of these projects make use of nonrepresentational approaches. Modeling tools involve the direct use of graphic manipulation to structure arguments, rather than simply providing a display. Nonrepresentational approaches use graphical means as primary instruments in production of arguments, but their relational and structural capacities can be used in display as well. The primary modeling function allows data structures and values to be generated from graphical manipulation. The components of a nonrepresentational system were described in a previous section. These features allow connections, organization (juxtaposition, sequencing, ordering, stacking, hierarchy) and other arrangements meant to be read semantically and rhetorically. The components of the system include layering, point of view systems, attributes (of completeness, uncertainty, ambiguity, and other features customizable by the user). The system was designed to assist in graphic display and inter-rogation/manipulation of elaborately marked-up narratives.[18] Temporal modeling provides an example of a project that makes similar use of graphi-cal elements and direct manipulation as acts of interpretation. In all cases, the system structure is the same: graphical features combine with elements that add affective value, meaning through relations, and other argument structures created directly. These all have conceptual dimensions and impli-cations. The graphical features are a set of elements with which to make direct interpretative actions using visual moves. Various features that draw on pictorial conventions, such as parallax and perspective, are incorporated for interpretative purposes. The graphical elements are culled from Jacques Bertin's foundational work on static graphics (size, shape, color, texture, posi-tion, orientation, and tone) with additional dynamic, animating elements that were not part of his system (change in size, position, orientation, speed, direction, etc.). Other elements would be borrowed from physics engines and animation software, such as bounce, repulsion, attraction, torque, weight, force, and so forth, to add features to the set of components to be used in interpretative activity. Though none of these features are meant to carry an inherent semantic value, they do conjure associations and can be used to create interpretative effects. Attributes can be used to signal qualitative value, even if they are given quantitative measure. These could be put at the ser-vice of intellectual schema, but also subjective assessments of pleasantness, anxiety, fear, distaste, and so on depending on the requirements of a project and the tasks of the scholar. The system includes activators and inflectors, or syntactic and semantic inflections. Activators are relational (such as contra-diction or similarity) and inflectors are largely semantic attributes (such as salience or ambiguity).

The dimensions of the system are meant to activate the potential of visual conventions in the service of interpretation. These include point of view, layers, slices, shifts in viewpoint such as tilt, splitting, folding, projecting, presentation from parallax views, and relative scales put into comparison. Annotation can be attached to any feature. The justification for these ele-ments is to allow a researcher to expose features of the data through graphic manipulation and thus to discover aspects of data otherwise unseen or hid-den in standard displays.

The distinguishing feature of nonrepresentational approaches to model-ing interpretation is that these reverse the standard sequence. Visualizations usually begin with data and then generate a visualization as an output using some regular convention such as a bar chart, line graphs, network diagram, and so on. But the nonrepresentational approach allows direct data production through graphical input and inflection. The direct manipulation also suggests that graphical organization, arrangement, manipulation, and features constitute arguments. The ordering of a set of documents, or the attraction of one piece of evidence to another, is a rhetorical move, and it creates an argument. A nonrepresentational argument space is a familiar feature of the analogue desktop where we arrange our papers for use, or the space of the now largely defunct slide table, where the spatialized organization of a lecture took place before being subjected to editing into linear presentation form. These aspects of analogue (and digital) manipulation have not been well integrated into digital humanities research, especially with the goal of making the graphical argument structure a primary method of generating data structures and values.

The projects share a few common features: they use direct methods of graphic activity (manipulating existing forms, objects, screen representations, or surrogates and creating new graphic forms, arrangements, or structures) to perform interpretative acts. These acts are specific to particular humanities research tasks—making timelines, chronologies, spatial representations, maps, networks, or arrangements of documents or other materials as part of an argument—but the tools can be used for tasks concerned with work in other domains (social sciences, policy, emergency management systems and so on). In every case, the graphic manipulation either *alters* an existing data structure, *creates* it, or *adds* to it. In every case, the graphic manipulation can be used to generate a nonstandard metric: a timeline that registers affective impact on relative scales, heterochronologies created according to culturally specific or historical models that do not match current standard western time-keeping, maps that are morphed and warped according to experience or evidence, models of partial knowledge either as percentages of a base reference image or as interconnected fragments constructing an image as an effect. All are premised on a nonrepresentational approach to graphical organization and expression as semantically meaningful as argument structures that make use of organization and arrangement in significant/signifying ways. All are designed to scale, but to work at a very granular, detailed level. All can import structured data and export newly structured data, but the complexity of data created in this approach (its inclusion of variables, nonstandard metrical features, and other elements generated within the visualizations) will make it difficult to export to or display in standard, conventional formats—but will add richness and dimensions to its analyses.

These projects and examples cover almost all instances of current conventions for what is commonly referred to as "information visualization" but should more properly be called "graphical expressions of interpretative processes." They are sketches or outlines for projects that would provide proof of concept of multiple aspects of nonrepresentational approaches to modeling interpretation. All intellectual inquiry that makes use of simple, static visualizations conceals assumptions, procedures, and decisions in representations (constructions) that masquerade as presentations (declarative statements). All visualizations in current use are reifications of misinformation. Epistemologically, this is not a situation that can be corrected, only exposed, by creating a graphical system that demonstrates the hermeneutics of intellectual work as constitutive of its objects of inquiry.

Experimental versions of these approaches are sketched in the appendix.

Appendix: Design Concepts and Prototypes

The concept of "nonrepresentational" visualizations is unfamiliar and even seems counterintuitive. How can an image *not represent* something? But the emphasis in a nonrepresentational approach is on *making* rather than *showing*. In other words, a nonrepresentational image is the first step in an analytic or interpretative process, not a display of the results of an analysis or interpretation. The point of this appendix is to make clear how this approach works—and looks.

For interpretative methods to achieve their full potential, they need to enter into data structures through the graphic space of production. Any image that can be drawn can be translated into a data structure. For images, this can be done in a variety of ways. An image can be stored as pixels (like a tapestry, with instructions for each point), as a vector (shape that can be manipulated), as a set of instructions for rendering (surfaces, light, textures), and so on. By contrast, information visualizations are usually expressions of preexisting data. They are designed to express quantities, relations among them, and to use display algorithms to show features of data like proximity relations, centrality, and other relative values. Adding features like nonstandard metrics into these processes is a matter of modifying the data structure, not just the display algorithm.

For example: Think of the difference between storing data that describes a rectangular bar (coordinates on a grid with standard x-y axes, and weight/color of line) versus a bar with one side deteriorating. The edge that is deteriorating *can* be described in terms of rate of breakdown of the line, area over which this takes place, and other factors. That is a data structure, not just a picture file. Likewise, a metric on which such a square appears can be altered—suppose the x axis is meant to show time passing and the deteriorating square is about a breakdown in diplomatic relations. The tick marks on the x axis might very well alter to indicate time pressure building, stretched moments of news cycles, compressed moments of tension, and stress indicators around decision points. All of these factors could be extracted through either computational analytics or traditional interpretative methods, or a combination of the two.

Figure 1 shows a standard metric charting a quantity on the *y* axis and a timeline on the *x* axis in the first panel (number of communications exchanged between two embassies in one day) and the effect of the changed information (or interpretation of information) in the next. The boundaries of the bar break because the timeframe is unclear and overlapping and communications are being exchanged so fast that it was not always evident even on which day they were received. The sequence of crossed communications cannot be sorted out. The effect is a time warp in which the day stretches, loses its boundaries, and the communications load is not fully calculable.

The difference between *making* and *showing* is embedded in the order in which the work is done. When I make a sketch of the events just described, I create a primary visualization rooted in *interpretation*. If I rely entirely on computational analytics and then make a visualization, I am showing the *results*, and I have not used the graphics as my primary argument space. If I then modify the visualization—adding features, changing the structure or relationships of entities on the screen, and so on—then I am *making* again. This iterative and interpretative process is not yet an aspect of visualization platforms linked to data structures.

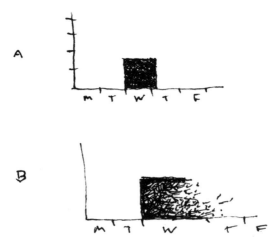

Figure 1
Standard metric (above) and interpretative metric (below), exposing the fallacy of discrete and bounded graphs. [JD]

This appendix provides visual documentation of projects through which I have explored the idea of modeling interpretation since my initial formulation of temporal modeling in 2000.[1] All of the projects share certain principles, such as the idea of affective metrics (metrics that are nonstandard) and direct construction of argument through visual means. These principles will be illustrated first, and then the six approaches outlined in chapter 5 will be described. These approaches contain features from various research projects, and so sometimes an image will be taken from Temporal Modeling, Ivanhoe, I.nterpret, or 3DH, the four different rubrics under which most of the work was undertaken. The visual materials here consist of sketches, drawings, some screen shots of prototypes, and other ways of envisioning the basic functionality and conceptual operations of modeling interpretation. The lack of a functioning platform may diminish the credibility of this work for some readers. But recent work in visualization shows an increased interest in creating interpretative, rather than empirical, approaches, and this research is meant to assist in that development.

The first group of images demonstrates principles that appear across the projects: graphic argument, generative (or affective) metrics, inflection, point-of-view systems, and direct input (the two-way screen). While data modification is part of the behavior supported by the projects, data display is still part of the basic functionality.

(1) Graphic argument

(2) Affective metrics

(3) Inflection

(4) Point-of view systems

(5) Direct input (two-way screen)

The second group of images shows ways to implement these principles in each of several major areas of visualization—temporal and spatial modeling, network diagrams, knowledge organization, document-based research and argument (aided by analytics), and interface. While not every type of visualization is shown here (tree maps, whisker diagrams, bubble charts, cluster diagrams, and so on are not addressed), the principles of affective metrics and graphic inflection could be used in any of these formats.

A. Temporal modeling and b. Heterochronologies (a version of the comparative ontologies)

B. Spatial modeling

C. Network inflection

D. Comparative ontologies

E. Argument creation and rich research field

F. Enunciative interface

Demonstration of principles

Graphic argument

The principle of *graphic argument* is fundamental to all nonrepresentational visualization. Graphic argument is the basic concept of using visual
means to create an argument structure. The graphical features of proximity,
sequence, ordering, placement, orientation, scale, and arrangement can all
carry semantic values. In an analog environment, laying out slides on a light
table, or organizing papers on a desk, or any other spatialized approach to
working with research materials performs this work. The advantage of the
digital environment is that this intuitively meaningful organization can be
translated into a data structure that is stored as a relational schema or model
independent of the contents of each object. Layers could be used to separate
different parts of the argument, as in figure 2.

Figure 3 shows three stages of argument development: top: display of
data sets; middle: argument space; and bottom: analytics generating visualizations that are in turn being manipulated to modify the data iteratively.

All of the graphic arguments depend upon the use of visual elements
in semantically meaningful, legible, and intuitive ways. The line between

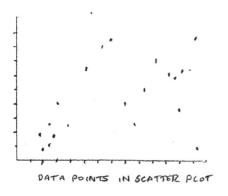
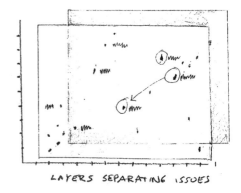

DATA POINTS IN SCATTER PLOT   LAYERS SEPARATING ISSUES

Figure 2
Layers used to separate data points and record graphical intervention. [JD]

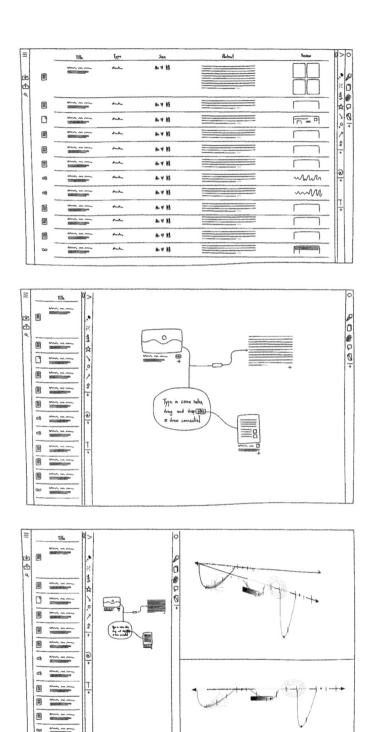

Figure 3
Sequence of argument development in graphical form. [Ibach]

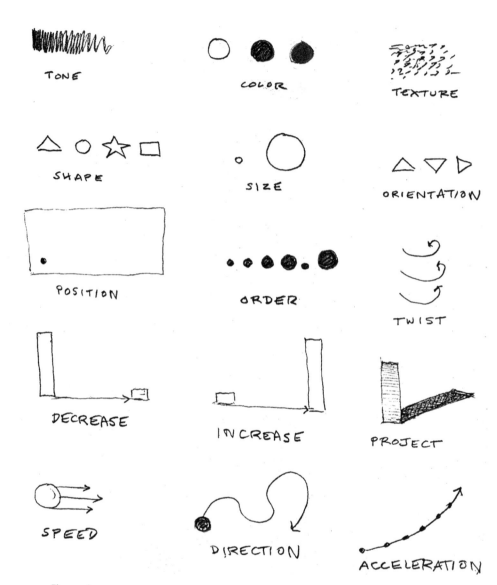

Figure 4
Graphic variables and inflections. [JD]

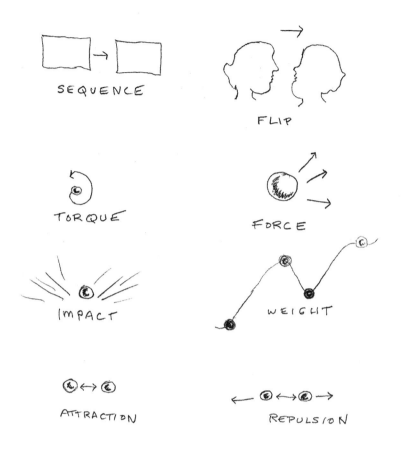

SEQUENCE

FLIP

TORQUE

FORCE

IMPACT

WEIGHT

ATTRACTION

REPULSION

USE AS ACTIVATORS / INFLECTORS:
CONTRADICTION, IMPORTANCE, RELEVANCE...
AMBIGUITY, UNCERTAINTY, SALIENCE....

Figure 4
continued

conventional uses of visual elements and intuitive ones can be unclear, but basic guidelines for use of graphic variables can be found in most interface and user design textbooks.

The image on the left in figure 4 shows basic graphic variables, borrowed from Jacques Bertin but augmented with animation elements and a few other features (like projection). Each can be assigned a unique role in a map, chart, interface, or design. Shapes, for instance, are easier to associate with different types of information while tonal values suggest relative degrees of intensity, and so on. The image on the right shows actions that can be taken to inflect or modify the presentation/creation of materials (documents or evidence, or components of a visualization).

Affective metrics

Affective or *generative* metrics are graphic expressions of nonstandard measure. These take into account, for instance, the effect of anxiety on the perception of time or space, or the impact of a particularly important piece of data on the structure of a graph or chart. In figure 5, the degree of anxiety has been tracked on the y axis against standard time intervals on the x axis. Then the length of lines, reflecting the rate of change, has been used to generate an x axis that incorporates these metrics. Any other phenomenon, such as expenditure of energy or analysis of sentiment, could be charted using the distortions introduced into the nonstandard metric, because these values are present in the experiential or user-dependent conditions being visualized.

Inflection

The concept of inflection is simply that of adding a graphic modification that nuances an otherwise blunt or reductive visual form. The capacity of human vision to register distinctions far exceeds the graphic vocabulary for charts and graphs currently in use. Inflection adds many subtleties and dimensions to the graphics. In figure 6, the difference between the images on the left and those on the right is dramatic. Graphical features like brightness, dullness, size, distance, and tonal definition carry semantic value by conventions that are easily legible.

In another example of inflection (figure 7), we can imagine that a standard representation of a quantity of data such as "how many residents approve of the proposed development" as a single bar on a chart does not do justice

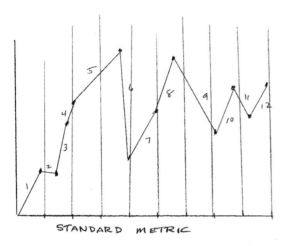

STANDARD METRIC

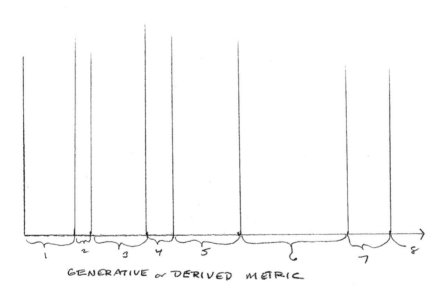

GENERATIVE or DERIVED METRIC

Figure 5
Generating an affective metric through systematic means. [JD]

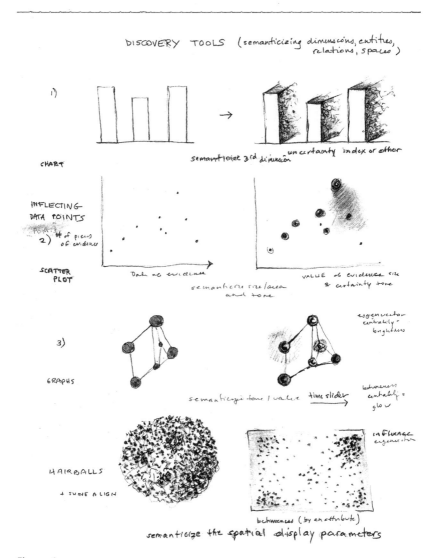

Figure 6

Graphical inflection applied to a network diagram. [JD]

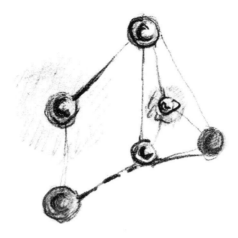

Figure 6
continued

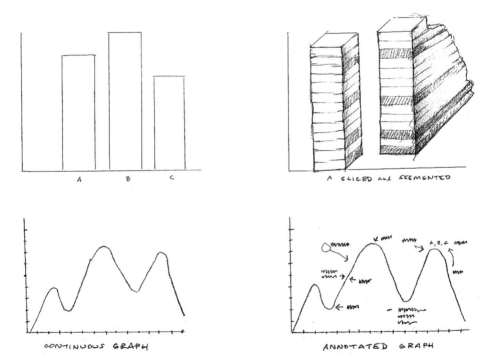

Figure 7
Slice technique to provide a view into the complexity of the data concealed in the
basic bar chart. [JD]

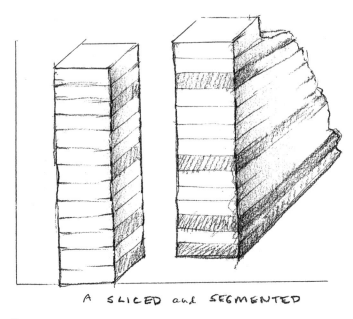

A SLICED and SEGMENTED

Figure 7
continued

to the complexity of the data set. A more nuanced presentation might show the specific makeup of that single bar in terms of other features in the data.

Point-of-view systems

Point of view inserts a subject position or user into the creation of an image. All images are created from some point of view, even if it is not marked. Perspectival images indicate the viewer/subject position through the creation of vanishing points that inscribe the position of a viewer within the structure of the image (as the place from which the image is drawn). If we take a standard scatter plot, the viewer is positioned as an omniscient, neutral subject seeing all points from the same position. But if we situate the viewer within the image, the relative values of the elements align in accord with a point-of-view system and vanishing points (figure 8). Imagine this as a chart of data displaying out-of-pocket medical insurance expenses (y axis) across time (x axis), first in an observer-neutral mode (left), then with respect to changing circumstances (now the z axis introduces a vanishing point). Then note the third rendering, in which a person in difficult

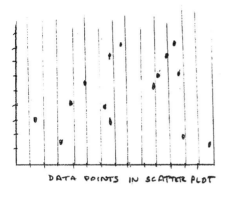

DATA POINTS IN SCATTER PLOT

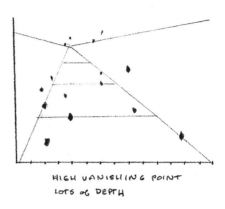

HIGH VANISHING POINT
LOTS of DEPTH

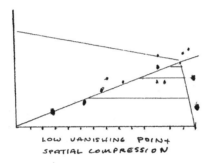

LOW VANISHING POINT
SPATIAL COMPRESSION

Figure 8
Point-of-view system introduced into a data visualization. [JD]

economic circumstances views the same data. The distortions indicate an affective relation to the information.

Direct input (two-way screen)

Like graphic argument, the two-way screen is an essential feature of nonrepresentational approaches to visualization and modeling (figure 9). It allows for direct manipulation through user input that *results* in a data set rather than simply giving the user a *display*. This is the feature that allows graphic argument to be implemented and to make an iterative loop between *making* and *showing*.

REPRESENTATION   MODELLING

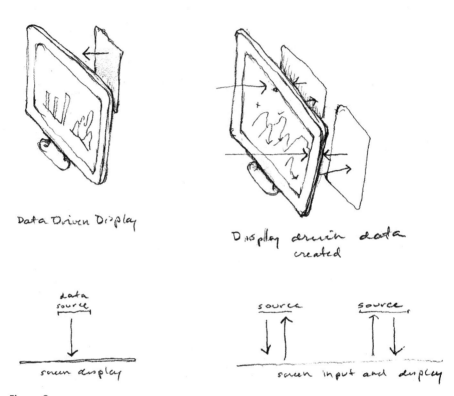

Figure 9
The two-way screen concept demonstrated, along with the production of information from graphical activity. [JD]

Projects

Having addressed a few basic principles of interpretative modeling, we can
now review their application within projects.

A. Temporal modeling

The temporal modeling project was created to address the conviction that
timelines designed for the empirical and social sciences are not suited to
work in the humanities. Empirical timelines are continuous, unidirectional,

and homogeneous in their metrics. In humanities documents, such as works of literature or film, the subjective experience of time *as well as* the presentation of temporal information is often discontinuous (gaps or jumps between one chapter or scene and another), multidirectional (flashbacks, reevaluation of past events in light of new evidence, and so on), and nonstandard (some minutes are longer than others). These principles apply to the *telling* (plane of *discourse* in linguistic terms or *syuzhet* in narratology) and the *told* (plane of *reference* and *fabula*). While *time* may be conceived of as a given, a container into which events or occurrences can be placed, *temporality* suggests a relational and experiential framework.

The primitives, or basic components, of the temporal modeling system were points, events, and intervals (figure 10)—a set of inflections divided into *semantic* and *syntactic*. *Semantic* inflections consisted of attributes attached to any single component. An event might be characterized as

Figure 10
Original temporal modeling screens. [Original screen shots redrawn by Peter Polack]

important or disturbing, or as relevant to one character and not another, and so on. But the attribute belonged to that single component. A *syntactic* inflection involved two or more components. These might be linked in a causal way, or in a reaction, or in any other codependent relation. Syntactic inflections could include anticipation, regret, and other effects of events on each other. The system contained a layering system to allow comparison among different models or parts of models. Finally, it included a now-slider, a slider that advanced the model along a separate timeline with tick marks outside of but correlated to the model. (This is the icon on the left in the lower gray menu bar that looks like an eye on a figure.) Missing from the implementation were a few crucial features, such as stretchy timelines and an elastic grid or base that could respond to affective values.

In figure 10, the temporal objects are displayed from the top menu, and so are the inflections, with semantic inflections in the dropdown menu in

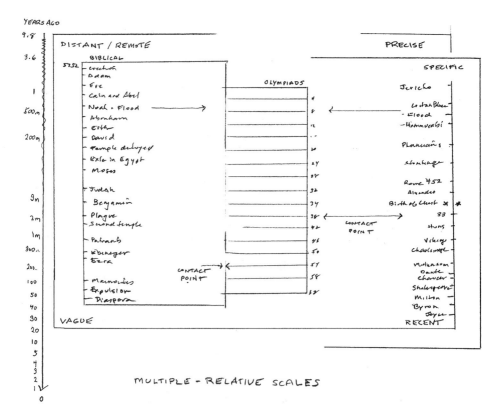

Figure 11
Heterochronologies and "floating" correlation points. [JD]

the middle and the syntactic ones displayed on the right. Color, glow, and labeling on objects and layers are shown below. These screen shots were from the implementation developed between 2000 and 2003.

B. Heterochronologies (a version of the comparative ontologies)

Heterochronologies put multiple chronological systems into comparison with each other. One vivid example is that of a biblical chronological framework and one based on geological time. They will have certain points of connection and will share certain dates and time periods, but not all. This is one example of comparative ontologies, when ontologies are understood as classification systems.

Figure 11 shows biblical, classical Greek, and geological timescales in relation to each other (at left), and includes a concept of vague versus precise temporal references (at right). Figure 12 shows a simple framework for charting temporal relations in two different dimensions—degree of

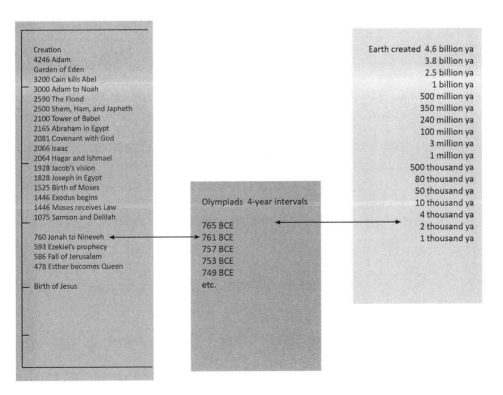

Figure 11
continued

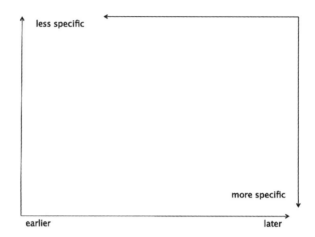

Figure 12
Conceptual framework for a relational model of chronological visualization. [JD]

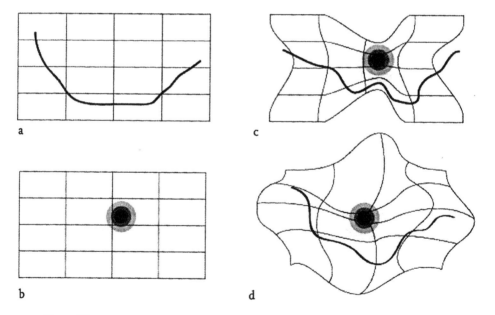

Figure 13
Spatial distortion as a result of affective charge. [Eskandar]

vagueness/specificity and relation to past/present position. This generalized scheme is a useful model in addressing the narrative of events in historical materials as well as fictional or aesthetic ones.

Spatial modeling

The concept of spatial modeling arises from the same impulse as temporal modeling—the belief that standard metrics are inadequate for showing how we experience spatial environments or for showing how these are presented in narrative and aesthetic work (or even journalistic work). The idea of affective mapping is demonstrated very simply in figure 13. A path appears on a regular cartographic grid in the first panel. In the second, an event occurs in the space. In the third panel, the space has contracted around the event, become compressed as a focal point, and the areas beyond the focus spread outward. This could be inverted, with the area where the event occurs actually expanding to show increased attention and detail (fourth panel).

In figure 14, the usual bird's-eye, observer-independent view of the famous map by Dr. Snow has been tilted so that the perspective situates the geographical events in a human framework.

Finally, figure 15 shows a map titled to position it within an experiential eyeline, and then the territory on the map is shown from two directions to demonstrate how parallax (differences in position of the sightline) changes perception of a detail of the landscape.

Network inflection

The concept of inflection has been discussed in relation to the temporal modeling project. It feels most important in the realm of network visualizations, where the assumptions built into the display suggest that all relations (edges) are the same and all nodes are separate from the edges and remain unaffected by changes in circumstances and conditions. As almost no relations among human beings, entities, or institutions are stable, this system would allow the changes to register, but also insists on a *nodedge* concept in which nodes and edges are codependent on, not independent of, each other. In figure 16 the nodes and edges both change.

Figure 17 shows the use of graphically applied inflection. The first panel shows a standard scatter plot; the second panel shows a line of connection being made between two data points that are also graphical objects on the screen; the third panel shows added inflection as an act of interpretation by

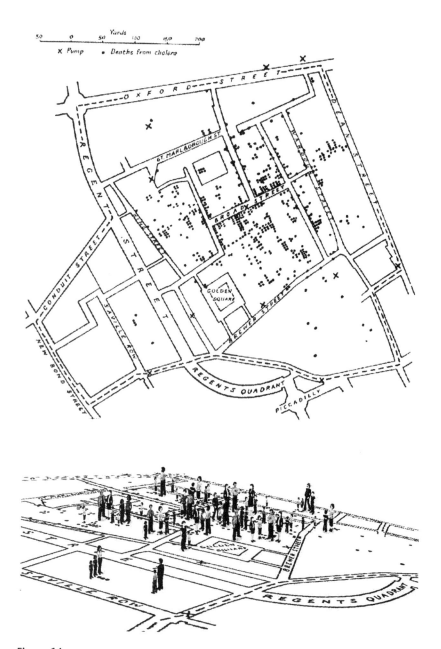

Figure 14
Point of view system and human identities positioned with Snow diagram. [Eskandar]

MAP
POINT OF VIEW

DISCOVERY TOOL

Figure 15
Parallax demonstrated on a cartographic representation. [JD]

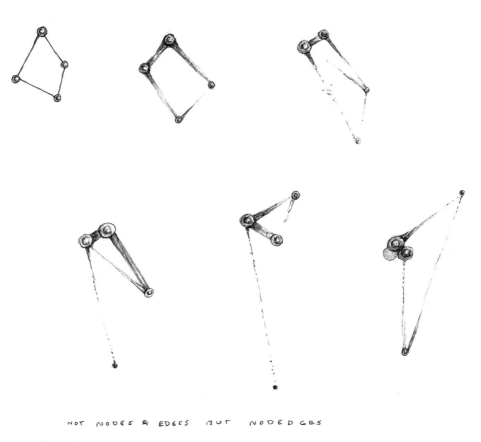

NOT NODES & EDGES BUT NODEDGES

Figure 16
Node and edge transformed into *nodedge* presentation. [JD]

a researcher. The data tables below include the transformed data structure
for each step.

Comparative ontologies

Comparative ontologies allow for contrast among different schemes of
knowledge organization. This approach has implications for cultural memory practices, and in particular for exposing various biases built into classification systems produced in, for instance, western, first-world museums,
archives, and libraries. Not all ontologies are structured in the same way.
Some are hierarchical, but some are flat, some are matrices, and some are
faceted. The purpose of this project is to make comparison among these systems visible. A feathered fan might, for instance, appear in a classification

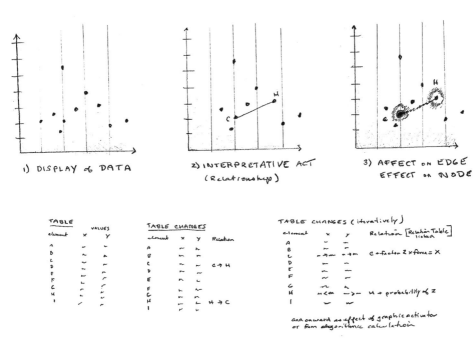

Figure 17
Generating complex data from direct graphical input. [JD]

system of fashion items, of ritual ethnographic pieces, or as evidence of an extinct species. Showing the relation among these and the extent to which an object belongs to multiple knowledge systems adds dimensions of cultural comparison into our visualizations.

Ontologies are usually expressed in language, even though they also depend upon graphic format to show hierarchy and groupings (sets and subsets). In figure 18 the object "feather fan" is shown in its place in multiple hierarchies. The identity of the fan, as an artifact, is complicated by these multiple schemes. Such schemes also inscribe worldviews and cosmologies, and the placement of objects within them is a nonneutral act whose specificity is made more clear by contrast.

Argument creation and a rich research field
One of the advantages of digital tools and computational analytics is the benefit of scale. Large corpora can be examined and searched, though caveats on any automated process are many, even as tools become more sophisticated about context-dependent meaning, inference, use of thesauri,

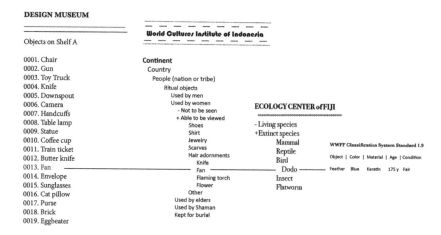

DESIGN MUSEUM

Objects on Shelf A

0001. Chair
0002. Gun
0003. Toy Truck
0004. Knife
0005. Downspout
0006. Camera
0007. Handcuffs
0008. Table lamp
0009. Statue
0010. Coffee cup
0011. Train ticket
0012. Butter knife
0013. Fan
0014. Envelope
0015. Sunglasses
0016. Cat pillow
0017. Purse
0018. Brick
0019. Eggbeater

World Cultures Institute of Indonesia

Continent
Country
People (nation or tribe)
Ritual objects
Used by men
Used by women
- Not to be seen
+ Able to be viewed
Shoes
Shirt
Jewelry
Scarves
Hair adornments
Knife
Fan
Flaming torch
Flower
Other
Used by elders
Used by Shaman
Kept for burial

ECOLOGY CENTER of FIJI

- Living species
+Extinct species
Mammal
Reptile
Bird
Dodo
Insect
Flatworm

WWFF Classification System Standard 1.9

Object | Color | Material | Age | Condition

Feather   Blue   Keratin   175 y   Fair

Figure 18
Comparative ontologies example in table form, floating correlation shown. [JD]

and entity recognition. But the relationships between humanities documents and analytic tools, as well as analog argument structures and data structure, have not been fully worked out. A good case study might be to think about how an area of land where a battle took place could be mapped using spatial modeling tools across a period of time in which certain crucial events take place (requiring stretched metrics) using various kinds of evidence (figure 19). The evidence might include letters, journal entries, weather reports, newspaper accounts, field notes, photographs, physical/ archaeological information, and other materials. The research would likely want to organize and order these materials, but also have a way to inflect the map with values generated from these documents. At an analytic level, processing meteorological data from a period of several months so that it can be correlated to events might be done best with computational tools. Sentiment analysis, also computationally generated, could be used to show the affective connection to the landscape and thus the importance of some areas more than others. In short, this platform/project encourages a back-and-forth iterative relation between traditional close reading, organization of evidence, argument construction, and a computationally supported generation of data used to model spatial features.

The full design of this research field would combine features of temporal and spatial modeling, inflection, comparative ontologies, and direct

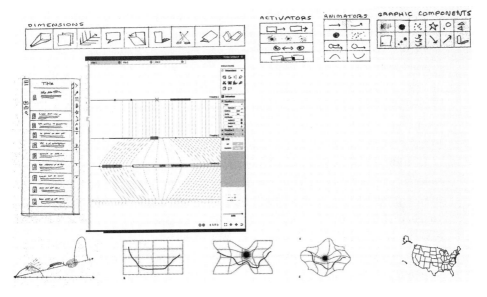

Figure 19
The rich research field correlating documents, timelines, chronologies, and cartographic expressions. [Ibach and JD]

input. It would synthesize elements of many of these examples into a single dashboard.

Enunciative interface

The challenge of creating a tool, platform, or visual convention for showing the "speaker" of an interface—or of parts of it—requires linking metadata from the internet feed and a display mechanism. When a webpage appears, it fills the screen window. No cracks or areas of unused screen real estate appear. And yet multiple "voices" are present. In a news site, the editorial and journalistic voices appear, sometimes unmarked by authorship attribution. Elements like mastheads or navigation systems are usually unattributed, while articles might have authorship indications. Advertisements are "spoken" quite differently than editorial material. Modes of direct address or indirect discourse may be marked in language, but not in the graphics.

Ways this might be done to break the illusion of seamless neutrality include "reveal codes" that show who is paying for each part of the display.

In addition, a legend could show whether the graphic work is a direct or oblique mode of address (looking at you, looking obliquely, or with back turned), situating the viewer in relation to the message and giving some indication of the demographic of the target audience. There could even be some amount of sentiment analysis indicating whether the content is within a norm for the site and how this compares with other sites (news treatments, for instance, could be visualized automatically to show spikes in tone or rhetorical deviation from their own norms or distance from each other).

In addition, a point-of-view system could be indicated in any web environment so that the site is shown through various filters or frameworks: from a point of view suitable for a six-year-old, a speaker of another language, a visually impaired user, or an adult with political leanings in one extreme or another. A filter device would make evident that any site is a constructed rather than neutral statement and that it is articulated by someone for some purpose.

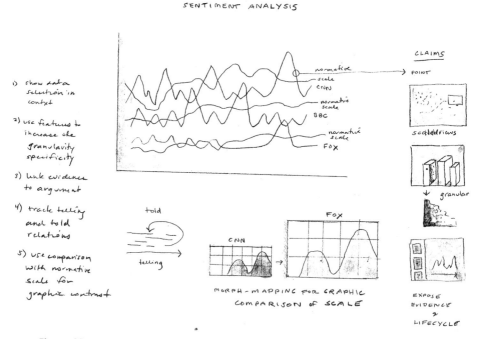

Figure 20

Site comparison to expose some of the enunciative features of an interface. [JD]

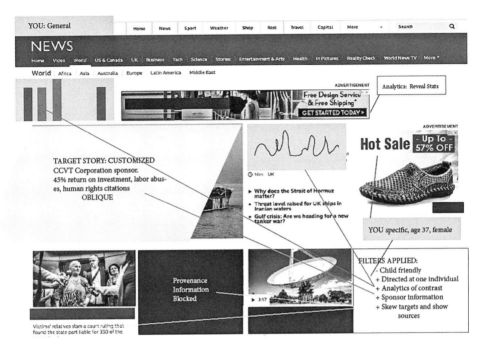

Figure 21
Modified interface to show enunciative features. [JD]

Figure 20 shows a model of site comparison using automation that could be embedded in a web browser. Figure 21 begins to sketch ways a combination of annotation, displacement, revelation, and referencing could alter the experience of an interface.

The drawings in figure 22 sketch a standard interface (completely flat/frontal) and two versions of a "popped-out" interface from different viewpoints. User profiles, analytics, and metadata trail/tags are all shown. The flat "naturalness" of the screen is altered by the point-of-view system, its alignments, its changes of scale, and its indexical connection to provenance information, sponsorship, and other usually invisible information in the metadata and in the analytics.

Other images and features

The gallery below contains ideas for other tools and techniques of manipulating visual display and/or visual image production as part of interpretative work. Most were developed for the 3DH project but not implemented.

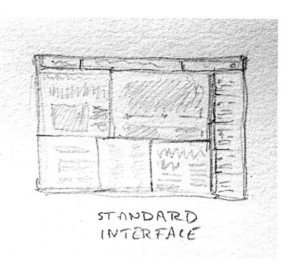

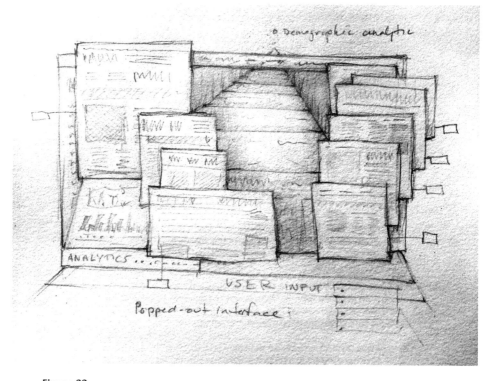

Figure 22

Standard interface and "popped-out" windows to expose enunciative features. [JD]

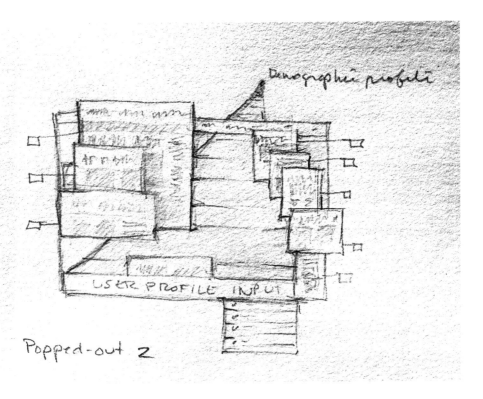

Demographic profile

USER PROFILE INPUT

Popped-out 2

Figure 22
continued

The 3DH system included research on using specific graphic features to indicate specific aspects of argument, such as salience, degrees of completeness or reliability, and ambiguity, such as those shown in figure 23. It also experimented with various conventions from architectural drawing and other modes, as per the palette of options in figure 24. Some of these have already been discussed, but others, like fold or point of view, are more novel. The explanation of function and concept are given below.

The idea of projection, or the use of a plane to intersect a view, is common in mathematics but not in data visualization. Figure 25 gives an idea of how distortion and anamorphosis could be used to show different views of data display. The second part of this figure shows how point of view can alter the presentation of a visualization. Imagine these as being generated with filters from opposed political positions.

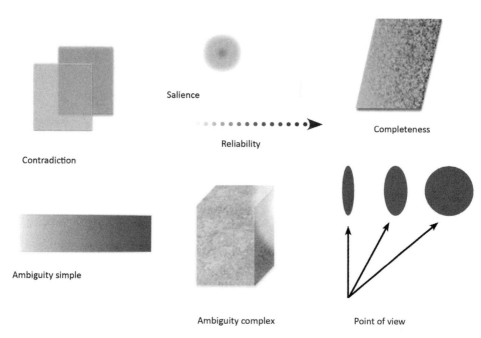

Contradiction

Salience

Reliability

Completeness

Ambiguity simple

Ambiguity complex

Point of view

Figure 23
Graphical features of the 3DH system. [JD]

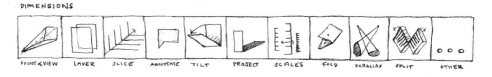

DIMENSIONS

POINT & VIEW   LAYER   SLICE   ANNOTATE   TILT   PROJECT   SCALES   FOLD   PARALLAX   SPLIT   OTHER

POINT OF VIEW: Author Attribute, Now slider, multiple values in contrast, use vanishing points; perspectivize to inhabit
LAYER: Partial knowledge, evidence, toggle values along slider
SLICE: Display on attribute, co-occurrence, any facet of data, track patterns
ANNOTATE: Add attributes to data, nodes, edges, txt, image; add cell/row or subelement
TILT: Skew along a line of bias or inquiry
PROJECT: Onto various planes according to angles of emphasis
SCALES: Relative scales kept distinct but with points of common contact
FOLD: Match on points and patterns to see discrepancies and alignments
PARALLAX: Multiple view lines into data, time models, maps, markup display
SPLIT: Cut in any dimension and view, slice, to see into, move, change scale of granularity
OTHER: Contrast, shift, see into, move, order, arrange, sequence

Figure 24
Dimensions of the 3DH system. [JD]

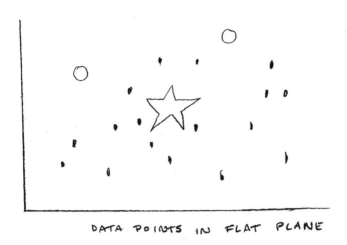

DATA POINTS IN FLAT PLANE

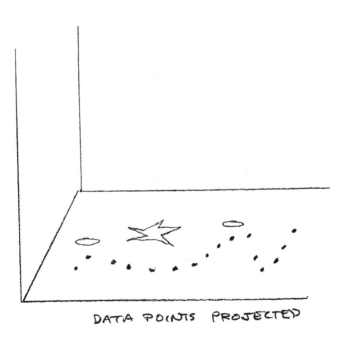

DATA POINTS PROJECTED

Figure 25
Projection and point of view as active features of a visualization environment. [JD]

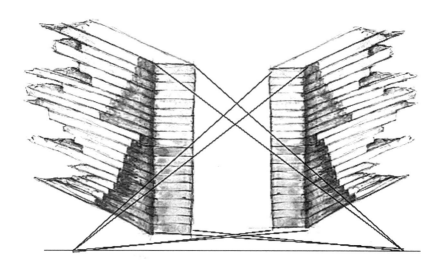

Figure 25
continued

Figure 26 shows data in a lifecycle of parameterization and display. This could be annotated with direct input as well.

The idea of taking the plane of a visualization and folding it at a crucial point to make comparisons is completely novel, but suggestive as a way of seeing differences in the development and continuation of situations. See figure 27, left side.

The right side of figure 27 uses a split or cut to carve off the portion of a visualization in which the maximum contrast occurs. A broken/implied axis is often used to show this when scale is preserved but the distance on the axis is so large that it is impractical to show the entire quantity being presented.

Figure 28 is a "heat map" approach to comparison. Any text or artifact can be humanly processed through markup or other means, or it can be processed in analytics using various parameters and filters. The comparison is in the layered rectangle in the middle of the image.

The density of data sets, their variation and granularity, is often ignored in favor of a reductive representation of a single quantity in a single metric scale. The approach here tries to "see into" the data at a higher level of specificity and also to suggest the nonuniformity of metric values (figure 29).

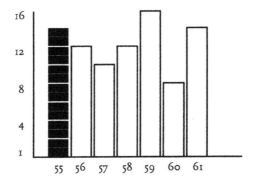

Fig 3a. A chart shows the number of new novels put into print by a single publisher in the years 1855-1862.

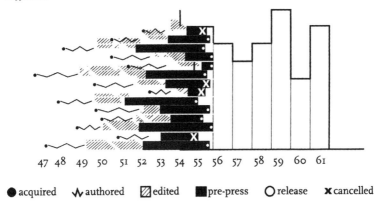

● acquired   ⋏⋏ authored   ▨ edited   ■ pre-press   ○ release   ✗ cancelled

Fig 3b. The "appearance" in 1855 of fourteen novels is shown in relation to the time of writing, acquisition, editing, pre-press work, and release thus showing publication date as a factor of many other processes whose temporal range is very varied. The date of a work, in terms of its cultural identity and relevance, can be considered in relation to any number of variables, not just the moment of its publication.

Figure 26
Reductive and complex presentations of information in visualization. [Eskandar]

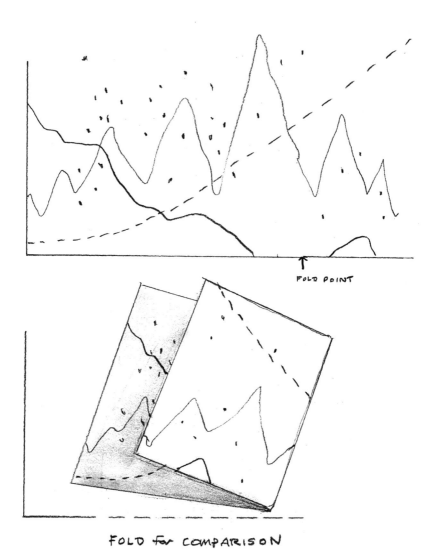

Figure 27
Innovative approaches to visual analysis: folding and slicing. [JD]

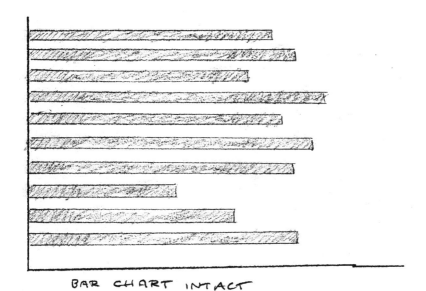

BAR CHART INTACT

SPLIT TO FOCUS ON DISCREPANCIES

Figure 27
continued

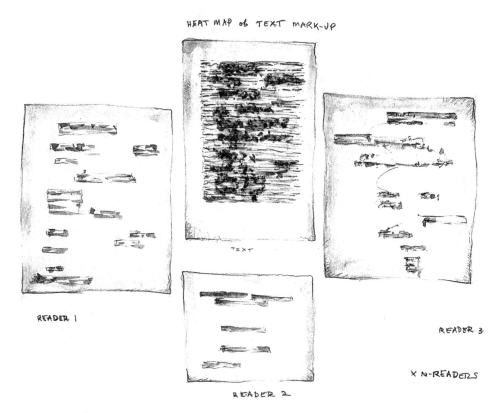

Figure 28
Heat map of text according to readings and markup. [JD]

Conclusion

The methods sketched here outline an agenda for an innovative approach
to modeling interpretation using graphical elements in connection with
more conventional analytics and display modes. No technical obstacles
prevent these from being developed in a working platform which could be
implemented at various scales of complexity, from simple graphic argument
to the more elaborate operating demands required for the rich research field
and its spatial and temporal modeling capacities.

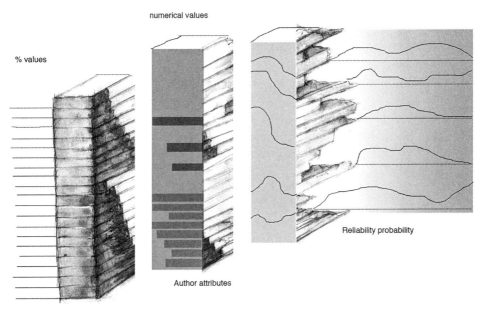

Figure 29
Split or cut-through of a data set to expose complexities. [JD]

The images in the appendix are either drawn by me [JD], or by Xárene Eskandar or Merle Ibach. Eskandar's work was done for hire, and Ibach's was done under terms of an agreement in which she granted rights of reproduction in exchange for acknowledgment. Peter Polack redrew the temporal modeling interface in figure 10.

Notes

Framework

1. Lisa Gitelman's edited volume, *"Raw Data" Is an Oxymoron* (Cambridge, MA: MIT Press, 2013), addresses these issues from within a critical digital media studies approach, but even the most straightforward introductory text, John Dillard's "5 Most Important Methods for Statistical Data Analysis" (n.d.) states very clearly what the pitfalls of each standard method are and introduces critical caveats on mean, standard deviation, regression, sample size, and other factors; see https://www.bigskyassociates.com/blog/bid/356764/5-Most-Important-Methods-For-Statistical-Data-Analysis (accessed August 19, 2018).

2. Calvin Schmid, *Statistical Graphics: Design Principles of Practices* (Brentwood, CA: John Wiley, 1983), and Stephen Few, *Show Me the Numbers: Designing Tables and Graphs to Enlighten* (El Dorado Hills, CA: Analytics Press, 2004), are two excellent sources for engagement with the basics of data visualization. Schmid's work is much older, written when statistical graphics were all print-based, but it contains solid, useful, pragmatic, and critically focused discussion of graphic techniques. Few's work is more recent, uses images generated entirely in Excel, and also addresses visualizations from a critical perspective. Both approach visualization from a positivist and empirical perspective without the least hint of critical, theoretical humanistic reflection.

3. The Cartesian system of *x* and *y* axes is highly rational, standardized, and conventional as a way to create an underlying grid structure for any visual images such as a graph, chart, or map.

4. The term critical hermeneutics refers to interpretative practices conceived within the tradition of Martin Heidegger and Hans-Georg Gadamer and continued into contemporary ramifications, see: Section 2.

5. Solomon Wiener, "History, Tradition, and Understanding in Gadamer's *Truth and Method*," *OR Journal* (April 2017), https://orintercollegiatejournal.com/2017/04/30/history-tradition-and-understanding-in-gadamers-truth-and-method1/; and Adrian

Costache, *Gadamer and the Question of Understanding: Between Heidegger and Derrida* (Lanham, MD: Lexington Books, 2016). See review by Lawrence K. Schmidt, https://ndpr.nd.edu/news/gadamer-and-the-question-of-understanding-between-heidegger-and-derrida-2/.

6. Work in the statistical analysis of reading comprehension is typical of this approach. For an example, see Cindy Brantmeier, "Statistical Procedures for Research on L2 Reading Comprehension: An Examination of ANOVA and Regression Models," *Reading in a Foreign Language* 16, no. 2 (October 2004), http://nflrc.hawaii.edu/rfl/October2004/brantmeier/brantmeier.html, and extensive research work in *Reading Research Quarterly*.

7. Those familiar with quantum theory will recognize the Copenhagen interpretation of Niels Bohr and Werner Heisenberg in this language. My use of quantum theory is not merely metaphoric, it is meant to describe the actual workings of interpretative methods and encounters. See Jan Faye, "Copenhagen Interpretation of Quantum Mechanics," July 24, 2014, *Stanford Encyclopedia of Philosophy*, https://plato.stanford.edu/entries/qm-copenhagen/.

8. See my *The Visible Word* (Chicago: University of Chicago Press, 1994), *The Alphabetic Labyrinth* (London: Thames and Hudson, 1994), *Figuring the Word* (New York: Granary, 1998), *SpecLab* (Chicago: University of Chicago Press, 2009), *Graphesis* (Cambridge, MA: Harvard University Press, 2014), *Diagrammatic Writing* (Eindhoven, Netherlands: Onomatopee, 2013), and many articles on related topics.

9. Concepts of the probabilistic relation of reader and text, and the idea of a text as a provocation for reading, were developed in conversation with Jerome McGann in the early 2000s in the period when we were creating the intellectual and design framework for Ivanhoe, a Game of Interpretation. The temporal modeling project preceded Ivanhoe and was designed in 2000–2001, while Ivanhoe followed in 2002–2003 as part of our SpecLab work. McGann's commitment to a probabilistic model, however, strongly resonated with the work I had done in some of the essays in *Figuring the Word* and the creative texts of *Deterring Discourse* (New York: Druckwerk, 1993), *Emerging Sentience* (New Haven: JAB books, 2001), and *Prove before Laying* (New Haven: Druckwerk, 1997). But my debt to McGann—and his to me—is evident in the shared vocabulary and conceptual overlap between his work in "Texts in N-Dimensions and Interpretation in a New Key [Discourse and Interpretation in N-Dimensions]," written in 2002. (See https://texttechnology.humanities.mcmaster.ca/pdf/vol12_2_02.pdf.) McGann's exposition in this piece is the most highly developed version of our thinking and surpasses my own in depth and detail. My appreciation of those conversations remains undiminished. However, missing from McGann's otherwise rich insights was any engagement with visuality. The digital humanities were almost entirely text-based in the 1990s, with very few exceptions.

Chapter 1

1. See Sarah Worth, "Art and Epistemology," https://www.iep.utm.edu/art-ep/, for evidence of this attitude and her arguments against it.

2. The term *iconography* refers to the meaning of visual images, particularly in their work as symbols.

3. Johanna Drucker, *Graphesis: Visual Forms of Knowledge Production* (Cambridge, MA: Harvard University Press, 2014).

4. Richard Gregory, *Eye and Brain: The Psychology of Seeing* (Princeton: Princeton University Press, 1996), is a classic, as are Rudolf Arnheim's, *Art and Visual Perception: A Psychology of the Creative Eye* (Berkeley: University of California Press, 1954) and *Visual Thinking* (London: Faber and Faber, 1969). See also Ernst Gombrich, *Art and Illusion* (Princeton: Princeton University Press, 1960).

5. Peter S. Wells, *How Ancient Europeans Saw the World* (Princeton: Princeton University Press, 2012), provides an anthropological approach; but for image analysis from within this conception, see John P. Snyder, *Flattening the Earth* (Chicago: University of Chicago Press, 1993), W. J. T. Mitchell, *The Last Dinosaur Book* (Chicago: University of Chicago Press, 1998), and Martin Rudwick, *Scenes from Deep Time* (Chicago: University of Chicago Press, 1995).

6. Harry N. Robin, *The Scientific Image: From Cave to Computer* (New York: Harry N. Abrams, 1992), is one good source for a treatment that includes analysis of the scientific issues in the images and their contribution. See also Michael E. Lynch and Steve Woolgar, eds., *Representation in Scientific Practice* (Cambridge, MA: MIT Press, 1990).

7. Wilfrid Blunt, with William T. Stearn, *The Art of Botanical Illustration: An Illustrated History* (London: Collins, 1950); Tim Lenoir, ed., *Inscribing Science: Scientific Texts and the Materiality of Communication* (Stanford: Stanford University Press, 1998); Lorraine Daston and Katharine Park, *Wonders and the Order of Nature 1150–1750* (New York: Zone Books, 2001); Bruno Latour, "Visualization and Cognition: Drawing Things Together," *Knowledge and Society* 6 (1986), 1–40; Karin Knorr-Cetina, *Epistemic Cultures: How the Sciences Make Knowledge* (Cambridge, MA: Harvard University Press, 1999); Alina Payne, ed., *Vision and Its Instruments* (College Park: Penn State University Press, 2014); David Freedberg, *Eye of the Lynx: Galileo, His Friends, and the Beginnings of Modern Natural History* (Chicago: University of Chicago Press, 2002).

8. Daniel Simons and Christopher Chabris created the original video to test "selective attention" in https://www.youtube.com/watch?v=vJG698U2Mvo; see also their jointly authored article, "Gorillas in Our Midst: Sustained Inattentional Blindness for Dynamic Events," *Perception* 28 (1999), 1059–1074.

9. Gombrich, *Art and Illusion*.

10. J. J. Gibson, *The Ecological Approach to Visual Perception* (Boston: Houghton Mifflin, 1986).

11. Jerome Lettvin, Humberto Maturana, Warren McCulloch, and Walter Pitts, "What the Frog's Eye Tells the Frog's Brain," *Proceedings of the IRE* 47, no. 11 (1959), is a much-cited and debated paper about "feature detectors" in the visual system.

12. Claudia Swan, "Illustrated Natural History," in Susan Dackerman, ed., *Prints and the Pursuit of Knowledge in Early Modern Europe* (Cambridge, MA: Harvard University Press, 2011).

13. Anton Stankowski, *The Visual Presentation of Invisible Processes: How to Illustrate Invisible Processes in Graphic Design* (New York: Hastings House, 1964).

14. Peter Galison, *Image and Logic: A Material Culture of Microphysics* (Chicago: Chicago University Press, 1997).

15. Eugene Ferguson, *Engineering and the Mind's Eye* (Cambridge, MA: MIT Press, 1992).

16. Denise Schmandt-Besserat, *How Writing Came About* (Austin: University of Texas Press, 1992), contains a discussion of the relations between ground lines, writing, and images.

17. I would be willing to make the case for any image on these terms—abstract formalism in the mode of Kandinsky, Ad Reinhardt, or Agnes Martin, collage and pastiche work, portraits and landscapes, small studies, vignettes, conceptual, procedural, and thematic work. Every image contains a presumption about the relation between vision and knowledge, perception and its social and cultural location and construction, and the contract between viewer and producer of the work.

18. The portrait being referenced is currently in the Staatliche Museen at Kassel

19. Rembrandt's evident preoccupation with this issue is evident in the nearly one hundred self-portraits he produced, with their varied answers to this question. Taken as a proposition, how can the subject know itself as an object, the formulation of self-portraiture is a continual investigation of the spectrum of beliefs according to which it can be answered.

20. Self-portraits are interesting in this regard, since they play with conventions of objectivity as well as positions of subjectivity across a wide range of conventions. Think of Dürer's self-presentation as a religious icon, or Klimt's as a figure of abjection.

21. For the notion of "pure difference," see Adam Nash, "An Aesthetics of Digital Virtual Environments," in Denise Doyle, ed., *New Opportunities for Artistic Practice in Virtual Worlds* (Hershey, PA: IGI Global, 2015), 7.

22. Garfield Benjamin, "'Virtual Reality' Reconsidered," in Drew Harrison, ed., *Handbook of Research on Digital Media and Creative Technologies* (Hershey, PA: IGI Global, 2015), 216.

23. Michael Eldred, "Digital Dissolution of Being," in *The Digital Cast of Being* (Frankfurt: Ontos Verlag, 2009), 2, https://www.scribd.com/document/59553025 /Digital-Dissolution-of-Being; see also Charlie Gere, "Digital Art and Visual Culture," in Ian Heywood and Barry Sandywell, eds., *The Handbook of Visual Culture* (London: Berg, 2012), 566: "In that it has no material existence and is based on pure difference the 'digital' is something like Derrida's Chora."

24. Alain Badiou, "The Ontology of Multiplicity: The Singleton of the Void" (2011), http://www.arasite.org/badiou2011a.html.

25. Johanna Drucker, "Digital Ontologies: The Ideality of Form in/and Code Storage—or—Can Graphesis Challenge Mathesis?," *Leonardo* 34, no. 2 (2001), 141.

26. Ibid.

27. Leibniz: "I thought again about my early plan of a new language or writing-system of reason, which could serve as a communication tool for all different nations" (cited by Dalakov, from a note, http://history-computer.com/Dreamers /Leibniz.html), and also "in his February, 1678, essay 'Lingua Generalis,' … connected closely with his binary calculus ideas[,] Leibniz spoke for his *lingua generalis* or *lingua universalis* as a universal language, aiming it as a lexicon of characters upon which the user might perform calculations that would yield true propositions automatically, and as a side-effect developing binary calculus" (ibid.).

"The only text in which Descartes explicitly mentions the expression *mathesis universalis* is in a passage from the *Rules for the Direction of the Mind*, where it is described as a 'general science that explains everything that it is possible to inquire into concerning order and measure, without restriction to any particular subject-matter.'" Frédéric de Buzon, "Mathesis Universalis," in Lawrence Nolan, ed., *The Cambridge Descartes Lexicon* (Cambridge: Cambridge University Press, 2015), https:// doi.org/10.1017/CBO9780511894695.166. The term does not show up anywhere else in Descartes, according to de Buzon.

See also J. Mittlestrass, "The Philosopher's Conception of *Mathesis Universalis* from Descartes to Leibniz," *Annals of Science* 36, no. 6 (1979), 563–610, https://www .tandfonline.com/doi/abs/10.1080/00033797900200401.

28. James Knowlson, *Universal Language Schemes in England and France 1600–1800* (Toronto: University of Toronto Press, 1975), as a start point.

29. Bishop John Wilkins, *An Essay Towards a Real Character, and a Philosophical Language* (London: Printed by John Martin for the Royal Society, 1668).

30. Erik Iversen, *The Myth of Egypt and Its Hieroglyphs in European Tradition* (Princeton: Princeton University Press, 1993).

31. Ernest Fenollosa and Ezra Pound, *The Chinese Written Character as a Medium for Poetry: A Critical Edition*, ed. Haun Saussy, Jonathan Stalling, and Lucas Klein (New York: Fordham University Press, 2008). The essay was originally published in 1919

and in spite of its inaccuracies had a great impact on modern poetry and Imagism in particular.

32. George Boole, *An Investigation into the Laws of Thought* (London: Walton and Maberley, 1854), and Ramon Cirera, ed., *Carnap and the Vienna Circle: Empiricism and Logical Syntax* (Amsterdam: Rodopi, 1994).

33. Norbert Wiener, *Cybernetics: or, Control and Communication in the Animal and the Machine*, 2nd ed. (1948; Cambridge, MA: MIT Press, 1961).

34. John Searle, "Cognitive Science and the Computer Metaphor," in B. Göranzon and Magnus Florin, eds., *Artificial Intelligence, Culture, and Language: On Education and Work* (Berlin-Heidelberg: Springer, 1990), 23–34, https://link.springer.com/chapter/10 .1007/978-1-4471-1729-2_4; Hubert Dreyfus, *What Computers Can't Do: The Limits of Artificial Intelligence* (New York: Harper and Row, 1972). See also work on the debates, Teodor Negru, "Intentionality and Background: Searle and Dreyfus against Classical AI Theory," *Filosofia Unisinos* 14, no. 1 (2013), http://revistas.unisinos.br/index.php /filosofia/article/view/fsu.2013.141.02; and Lee Gomes, "When Computers Are Not Really 'Brains,'" *Forbes*, April 23, 2009, https://www.forbes.com/forbes/2009/0511/046 -artificial-intelligence-neuroscience-digital-tools.html#4f64ee46b455.

35. John Searle, "Can Computers Think?," in *Minds, Brains, and Science* (Cambridge, MA: Harvard University Press, 1983), 28–41.

36. William Ivins, Jr., *Prints and Visual Communication* (Cambridge, MA: Harvard University Press, 1953).

37. Nelson Goodman, *Languages of Art: An Approach to a Theory of Symbols* (Indianapolis: Bobbs-Merrill, 1968).

38. Estelle Jussim, *Visual Communication and the Graphic Arts* (New York: R. R. Bowker, 1974); the reference work of Bamber Gascoigne, *How to Identify Prints: A Complete Guide to Manual and Mechanical Processes from Woodcut to Inkjet* (London: Thames and Hudson, 1986); and also the extremely valuable Graphics Atlas at Rochester Institute of Technology, http://www.graphicsatlas.org/.

39. René Thom, "Stop Chance! Stop Noise!," *SubStance*, no. 40 (1982), 9–21, special issue on OuLiPo.

40. Drucker, *Graphesis*, 23, contains another version of this argument and citation. Other references to Thom that foreshadow the version here follow on pp. 24–25. An early version of some of these arguments appeared as "Graphesis" in *PAJ (Poetess Archive Journal)*, special issue, "Visualizing the Archive," March 2011, http://tei-l .970651.n3.nabble.com/Visualizing-the-Archive-td2657423.html.

41. Martin Jay, *Downcast Eyes* (Berkeley: University of California Press, 1993).

42. The attention of structuralist critics and semioticians produced an exhaustive study of these issues and contrasts, beginning with Roman Jakobson, "On Some

differences," and proceeding through the work of Norman Bryson, Umberto Eco, Mieke Bal, Roland Barthes, and numerous others.

43. See Drucker, *Graphesis*, for a long discussion of these formal approaches within late nineteenth-century and twentieth-century modernism.

44. Jacques Bertin, *The Semiology of Graphics: Diagrams, Networks, Maps* (Madison: University of Wisconsin Press, 1983), originally published in French in 1967 as *Sémiologie graphique* (The Hague: Mouton; Paris: Gauthier-Villars, 1967).

45. This section of the chapter draws on material from Drucker, "Digital Ontologies," 141–145.

46. Matthew Kirschenbaum, *Mechanisms: New Media and the Forensic Imagination* (Cambridge, MA: MIT Press, 2008), especially chapter 1, "Every Contact Leaves a Trace."

47. Hubertus von Amelunxen, Stefan Iglhaut, Florian Rötzer, et al., eds., *Photography after Photography: Memory and Representation in the Digital Age* (Amsterdam: G+B Arts, 1996).

48. In Drucker, "Digital Ontologies," this part of the discussion focused on photographic images, not data representations, as well as on images generated from algorithms, such as the early pioneering work of Jack Citron and George Nees from the 1970s.

49. Karen Barad, *Meeting the Universe Halfway: Quantum Physics and the Entanglement of Matter and Meaning* (Durham: Duke University Press, 2007).

Chapter 2

1. Gerald Bruns, *Hermeneutics: Ancient and Modern* (New Haven: Yale University Press, 1992).

2. Ibid., "Introduction."

3. Jerome McGann, "Texts in N-Dimensions," *Text Technology* 12, no. 2 (2003), 10, http://texttechnology.mcmaster.ca/pdf/vol12_2_02.pdf; McGann, *The Scholar's Art: Literary Studies in a Managed World* (Chicago: University of Chicago Press, 2006); and the earlier work, *Radiant Textuality: Literature after the World Wide Web* (New York: Palgrave, 2001), though it does not have any discussion of Ivanhoe.

4. Canadian poet-theorists Steve McCaffery, Christian Bök, Darren Wershler, and the members of the Collège de Pataphysique, among others, have also used Jarry's work as a foundation for critical practice.

5. SpecLab was a loose association of individuals of whom some were graduate students at the University of Virginia paid for their research, one was a paid professional, and others were faculty colleagues, but it was never a formal "lab" in

any institutional sense. Among the individuals who worked with us were Bethany Nowviskie, Andrea Laue, Nick Laiacona, Worthy Martin, Geoffrey Rockwell, Steve Ramsay, and others. Credit goes to all of them in varying degrees for their contributions to the projects McGann and I envisioned.

6. Quite a few versions of the history of early digital humanities exist, but Susan Hockey, "The History of Humanities Computing," is a reliable introduction: http:// www.digitalhumanities.org/companion/view?docId=blackwell/9781405103213 /9781405103213.xml&chunk.id=ss1-2-1&toc.depth=1&toc.id=ss1-2-1&brand=default. Her *Electronic Texts in the Humanities: Principles and Practice* (Oxford: Oxford University Press, 2001) is still a useful reference work.

7. The problem of overlapping hierarchies was a major point of conversation in the 1990s and early 2000s. The question of whether documents, particularly literary and aesthetic ones, were themselves "ordered hierarchies of content objects" or merely being forced into conformance with such a model for the purposes of being "marked-up" with XML tags, produced raging debates. The arguments against the OHCO thesis are so evidently correct that it now seems odd to imagine there was such resistance, but the requirement for markup to follow nested hierarchies put much at stake in the discussion—especially as no evident solution could be created to deal with even such simple conflicts as the fact that a poem might begin and end on different pages while being constituted as a single work, or that the figurative or metaphoric use of a word in a poem would link it to multiple hierarchies that were not nested in the least. These simple examples are so fundamental to working with literary documents that the idea that scholars shrugged, gave in, and used an OCHO model now seems inconceivable.

8. The discussion here draws on Johanna Drucker, "Humanistic Theory and Digital Scholarship," in Matthew Gold and Lauren Klein, eds., *Debates in Digital Humanities* (Minneapolis: University of Minnesota Press, 2012), 88–89.

9. See Andrew Robichaud and Cameron Blevins, "Tooling Up for Digital Humanities," Stanford University, for a good introduction to many of these topics and tools for their implementation: http://toolingup.stanford.edu/?page_id=201.

10. Johanna Drucker, "Humanities Approaches to Graphical Display," *DHQ* (*Digital Humanities Quarterly*) 5, no. 1 (March 2011). This piece contains a formulation of the "data" and "capta" distinction applied to visualization. Bits of that article are paraphrased throughout.

11. McGann, "Texts in N-dimensions," 18. The reference is to Fish's essay "Is There a Text in This Class?" in which Fish examines that question after it is posed to him by a prospective student. Fish reflects on all of the many contextual frames required to actually "read" the apparently simple question and respond to it appropriately.

12. A useful overview is Marianne Freiberger, "A Ridiculously Short Introduction to Some Very Basic Quantum Mechanics," *+Plus Magazine*, May 2016, https://plus .maths.org/content/ridiculously-brief-introduction-quantum-mechanics.

13. Ernst von Glasersfeld, *Radical Constructivism: A Way of Knowing and Learning* (Washington, DC: Falmer Press, 1995), is a key text here. In addition, work by Humberto Maturana and Francisco G. Varela, *Autopoiesis and Cognition: The Realization of the Living* (Dordrecht: D. Reidel, 1980) and *The Tree of Knowledge: The Biological Roots of Human Understanding* (Boston: New Science Library, 1987).

14. Laura Mandell, "Gendering Digital Literary History: What Counts for Digital Humanities," in Susan Schreibman, Ray Siemens, and John Unsworth, eds., *A New Companion to Digital Humanities* (Chichester, UK: Wiley Blackwell, 2016).

15. As mentioned above, the team included Bethany Nowviskie, Nick Laiacona, and others who helped implement and test the design features and create the functional prototype. See Johanna Drucker, "Designing Ivanhoe," *Text Technology*, no. 2 (2003), http://www.t5d.com/tt/2004/pdf/vol12_2_03.pdf.

16. Ibid.

17. Ibid.

18. The discussion here draws on Johanna Drucker, "Performative Materiality and Theoretical Approaches to Interface," *DHQ* (*Digital Humanities Quarterly*) 7, no. 1 (Summer 2013), http://www.digitalhumanities.org/dhq/vol/7/1/000143/000143 .html, and Drucker, "Humanities Approaches to Graphical Display."

19. Again, the phrases here and basic principles rhyme with those of McGann, "Texts in N-Dimensions," for obvious reasons. Our work was completely meshed in the conversations about Ivanhoe and our vocabulary overlapped.

20. An excellent introduction to these principles is Philip Yaffe, "Making Sense of Nonsense: Writing Advice from Lewis Carroll and the Jabberwocky," *Ubiquity*, May 2008, https://ubiquity.acm.org/article.cfm?id=1386855; for more on Turing and the Love Letter Generator, see Noah Wardrip-Fruin, "Digital Media Archaeologies: Interpreting Computational Processes," in Erkki Huhtamo and Jussi Parikka, eds., *Media Archaeology: Approaches, Applications, and Implications* (Berkeley: University of California Press, 2011), 302–322; also Homay Kay, "Alan Turing's Automated Love Letter Generator," "https://dukeupress.wordpress.com/2014/12/01/4701/.

21. The discussion here draws on Drucker, "Humanistic Theory and Digital Scholarship," 86–87.

22. Patrik Svensson, in his experiments at HumLab and after, has frequently worked with multiple screens and sites to avoid this kind of monocular perspective.

23. Mandell, "Gendering Digital Literary History."

24. See the discussion in Drucker, "Humanities Approaches to Graphical Display."

25. McGann, "Texts in N-Dimensions," 10.

Chapter 3

1. Edward Tufte, *The Visualization of Quantitative Information* (Chesire, CT: Graphics Press, 1983).

2. J. J. O'Connor and E. F. Robertson, "A History of Topology," http://mathshistory .st-andrews.ac.uk/HistTopics/Topology_in_mathematics.html (November 12, 2004), or Teo Paoletti, "Leonard [sic] Euler's Solution to the Konigsberg Bridge Problem," Mathematical Association of America (August 21, 2018), https://www.maa.org/press /periodicals/convergence/leonard-eulers-solution-to-the-konigsberg-bridge-problem.

3. Johann Benedict Listing, biography; http://www-history.mcs.st-and.ac.uk/Biogra phies/Listing.html.

4. Georgi Dalakov, "Sketchpad of Ivan Sutherland," *History of Computers*, http:// history-computer.com/ModernComputer/Software/Sketchpad.html.

5. Dalakov (ibid.) on Ivan Sutherland: "Sketchpad stores explicit information about the topology of a drawing. If the user moves one vertex of a polygon, both adjacent sides will be moved. If the user moves a symbol, all lines attached to that symbol will automatically move to stay attached to it. The topological connections of the drawing are automatically indicated by the user as he sketches. Since Sketchpad is able to accept topological information from a human being in a picture language perfectly natural to the human, it can be used as an input program for computation programs which require topological data, e.g., circuit simulators."

6. On Neolithic circular enclosures in Central Europe, https://en.wikipedia.org/wiki /Neolithic_circular_enclosures_in_Central_Europe.

7. The balance of this chapter draws on material from Johanna Drucker, "Non-representational Approaches to Modelling Interpretation in a Graphical Environment," *Digital Scholarship in the Humanities* 33, no. 2 (June 2018).

8. Rhino is an example of rendering software, for instance, and though an argument can be made that the creation of a highly resolved image of an object is a model, even, on some level, an epistemological argument, it just confuses matters to consider drawing platforms and intellectual modeling as doing the same work.

9. Mind mapping software is often brought up as an example of this approach, but again, the tools are reductive, the ground on which the figures are drawn is standardized, Cartesian, and inadequate to express discontinuity, affective metrics, and other features of inflected experience.

10. Nigel Thrift, *Non-representational Theory: Space, Politics, Affect* (London: Routledge, 2008).

11. Jacques Bertin, *Semiology of Graphics* (Madison: University of Wisconsin Press, 1983); Leland Wilkinson, *The Grammar of Graphics* (New York: Springer, 2005).

Chapter 4

1. Brenda Laurel, ed., *Art of Human-Computer Interface Design* (Reading, MA: Addison-Wesley, 1990); Lori Spencer, *Reading Writing Interfaces* (Minneapolis: University of Minnesota Press, 2014); Ben Shneiderman, *Designing the User Interface: Strategies for Effective Human-Computer Interaction* (Reading, MA: Addison Wesley, 1997); Alex Galloway, *The Interface Effect* (Cambridge, UK: Polity Press, 2012); Matthew Fuller, *Behind the Blip: Essays on the Culture of Software* (Brooklyn, NY: Autonomedia, 2003); Steven Johnson, *Interface Culture: The Way We Create and Communicate* (New York: Harper Edge, 1997).

2. See Johanna Drucker, "Humanities Approaches to Interface Theory," *Culture Machine* 12 (2011), 2.

3. Ibid., 3.

4. Jeff Han at TED, https://www.ted.com/speakers/jeff_han.

5. Hayden Walles, "A Keyboard in the Palm of Your Hand," *Sydney Morning Herald*, December 1, 2011, https://www.smh.com.au/technology/a-keyboard-in-the-palm-of -your-hand-20111129-1o4ut.html.

6. Don Norman, *The Design of Everyday Things* (New York: Basic Books, 1990), and other work by Norman; Sachin Rekhi, "Don Norman's Principles of Interaction Design," *Medium*, January 23, 2017, https://medium.com/@sachinrekhi/don -normans-principles-of-interaction-design-51025a2c0f33; Pau Alsina, "PIPES Interface —User, Subjectivation, Materiality" (2015), https://interfacemanifesto.hangar.org /index.php/PIPES_Interface_%E2%80%93_User,_Subjectivation,_Materiality_-_Notes _By_Pau_Alsina, follows the outline, issues, and specific references of my argument, but includes some useful additions.

7. Mukurtu, http://mukurtu.org/.

8. The discussion here draws on Drucker, "Humanities Approaches to Interface Theory," 1.

9. This observation is based on using Google nGram to track the term. It does not have any apparent use before 1920, then climbs slowly with a rapid rise starting in the 1970s.

10. Alsina, "PIPES Interface."

11. For an overview of Engelbart's contribution, see https://www.dougengelbart.org /content/view162/000.

12. For basic guidelines from the United States Government, see https://www .usability.gov/what-and-why/user-interface-design.html.

13. As far back as the 1990s, concern about this topic surfaced among psychologists; see Kimberly S. Young, "Internet Addiction: The Emergence of a New Clinical Disorder," *CyberPsychology and Behavior* 1, no. 3 (1998), 237–244.

14. This and the following section draw on Drucker, "Humanities Approaches to Interface Theory."

15. Roger Chartier, "Languages, Books, and Reading from the Printed Word to the Digital Text," *Critical Inquiry* 31, no. 1 (Autumn 2004), 133–152. See also Johanna Drucker, "The Virtual Codex: From Page Space to eSpace," *Philobiblon* (2003), http://www.philobiblon.com/drucker/; Johanna Drucker, "From Entity to Event: From Literal, Mechanistic Materiality to Probabilistic Materiality," *Parallax* 15, no. 4 (2009), 7–17.

16. Geoffrey Nunberg, "The Places of Books in the Age of Electronic Reproduction," *Representations* (1993); Roger Chartier, *L'ordre de livres* (Aix-en-Provence, 1992), 20.

17. Brenda Laurel, *Computers as Theatre* (Reading, MA: Addison-Wesley, 1991).

18. Galloway, *The Interface Effect*.

19. Patrick Jagoda on Alex Galloway, *LARB*, https://lareviewofbooks.org/article/the-next-level-alexander-r-galloways-the-interface-effect/#!.

20. Jesse James Garrett, *The Elements of User Experience* (San Francisco: New Riders, 2002), and the much-published image, http://www.jjg.net/elements/.

21. Aaron Marcus and Associates, "Cross-Cultural User Experience Design," 2005, http://www-inst.eecs.berkeley.edu/~cs160/sp06/lectures/lec21/lec21.pdf.

22. Mauro Marinilli, "The Theory behind User Interface Design, Part One," 2002, developer.com, https://www.developer.com/design/article.php/1545991/The-Theory-Behind-User-Interface-Design-Part-One.htm.

23. Electronic Literature Organization, https://eliterature.org/.

24. Drucker, "Humanities Approaches to Interface Theory," 1 and 13; Ben Shneiderman, "Eight Golden Rules of Interface Design," https://www.interaction-design.org/literature/article/shneiderman-s-eight-golden-rules-will-help-you-design-better-interfaces.

25. A google search on HCI will turn up a host of examples of firms promoting their work with this rhetoric. See also the very funny Steve Krug, *Don't Make Me Think: A Common Sense Approach to Web Usability* (San Francisco: New Riders, 2000), for a realistic critical approach to interface design.

26. Kaja Silverman, *The Subject of Semiotics* (New York: Oxford University Press, 1983); Paul Smith, *Discerning the Subject* (Minneapolis: University of Minnesota Press, 1988); Stephen Heath and Teresa de Lauretis, eds., *Cinematic Apparatus* (New York: St. Martin's Press, 1980); Laura Mulvey, "Visual Pleasure and Narrative Cinema," in Leo Braudy and Marshall Cohen, eds., *Film Theory and Criticism* (New York: Oxford University Press, 1999), 833–844; Margaret Morse, *Virtualities: Television, Media Art and Cyberculture* (Bloomington: Indiana University Press, 1998).

27. Erving Goffman, *Frame Analysis* (Cambridge, MA: Harvard University Press, 1974).

28. Donald Hoffman, http://www.cogsci.uci.edu/~ddhoff/interface.pdf; Donald Hoffman, "The Interface Theory of Perception," in Sven J. Dickinson, Aleš Leonardis, Bernt Schiele, and Michael J. Tarr, eds., *Object Categorization: Computer and Human Vision Perspectives* (Cambridge: Cambridge University Press, 1989), http://www.veronadesign.biz/interface.pdf.

29. This section draws on Drucker, "Humanities Approaches to Interface Theory," 15; and on Johanna Drucker, "Performative Materiality and Theoretical Approaches to Interface," *DHQ* (*Digital Humanities Quarterly*) 7, no. 1 (Summer 2013), http://www.digitalhumanities.org/dhq/vol/7/1/000143/000143.html.

30. Émile Benveniste, "The Nature of Pronouns," in Benveniste, *Problems in General Linguistics*, trans. Mary Elizabeth Meek (1966; Coral Gables, FL: University of Miami Press, 1971).

31. This and the following section draw on Johanna Drucker, "Information Visualization and/as Enunciation," *Journal of Documentation* 73, no. 5 (2017), 903–916, https://doi.org/10.1108/JD-01-2017-0004.

32. John Phillips, "Who Is the Subject of Enunciation?," 2006, https://courses.nus.edu.sg/course/elljwp/enunciation.htm.

33. Bernd Frohmann, "Discourse Analysis as a Research Method in Library and Information Science," *Library and Information Science Research* 16 (1994), 133, 134; Bernd Frohmann, "The Power of Images: A Discourse Analysis of the Cognitive Viewpoint," *Journal of Documentation* 48 (1992), 365–386; Mark Poster, "Databases as Discourse," in E. Zureik and D. Lyon, eds., *Computers, Surveillance, and Privacy* (Minneapolis: University of Minnesota Press, 1996); Michael Strangelove, "Poster on Database Discourse," *Foucault Info*, April 8, 1997, https://foucault.info/foucault-l/msg02993.html.

34. See Benveniste, "The Nature of Pronouns" (1966), 224–225, cited in Drucker, "Information Visualization and/as Enunciation."

35. See Johanna Drucker, "Design Agency," *Dialectic* 1, no. 2 (2017), 11–16.

Chapter 5

1. Concepts like "the spatial turn" assigned to trends in the digital humanities, for instance, are generally referring to an uptake in use of maps, rather than a serious engagement with issues of space and spatiality, their conception across historical and cultural domains, or critical reflection on the definition of these terms.

2. Excel and Google Sheets are two outstanding examples, both of which enable a wide variety of visualizations. In *Show Me the Numbers* (El Dorado Hills, CA: Analytics Press, 2012), Stephen Few demonstrated the versatility and range of Excel's capabilities by making all of the visualizations in his book within its constraints.

3. See Johanna Drucker, *SpecLab* (Chicago: University of Chicago Press, 2009), 37–64, for a description of this project and its development. Temporal modeling was the first project I initiated at the University of Virginia in 1999–2000, and the founding of SpecLab and work on Ivanhoe picked up many of the features in that project in the summer of 2000 and after.

4. John David Miller and John Maeda, "A Stitch in Time: Visualizing History through Unit Forms and Repetition Structures" (2015), https://www.researchgate .net/publication/277250414_A_Stitch_in_Time_Visualizing_History_Through_Unit _Forms_and_Repetition_Structures.

5. J. T. Fraser, *Time, the Familiar Stranger* (Amherst: University of Massachusetts Press, 1987); and J. T. Fraser, Marlene P. Soulsby, and Alexander Argyros, eds., *Time, Order, Chaos: The Study of Time* (Madison, CT: International Universities Press, 1998).

6. See Drucker, *SpecLab*, 48, and 50–51, for specifics.

7. Working with a small team (Maura Tarnoff, Bethany Novwiskie, and Jim Allman, as well as Louise Sandhaus and a group of her students at CalArts).

8. In essence, this is what the individual viewpoints in Ivanhoe, in combination with the log and now-slider, permitted. See Johanna Drucker, "Designing Ivanhoe," *Text Technology*, no. 2 (2003), http://www.t5d.com/tt/2004/pdf/vol12_2_03.pdf.

9. Martin J. S. Rudwick, *Scenes from Deep Time: Early Pictorial Representations of the Prehistoric World* (Chicago: University of Chicago Press, 1992); for a description of the discovery of "Deep Time," see M. Alan Kazlev's site: http://palaeos.com /timescale/historical/index.html.

10. See Robert D. Montoya and Seth R. Erickson, "Anachronism in Global Information Systems: The Cases of Catalogue of Life and Unicode," https://www.ideals .illinois.edu/bitstream/handle/2142/98866/2pt13_Montoya-Anachronism.pdf ?sequence=1.

11. *Die Fackel* online, https://archive.org/details/diefackel64krauuoft.

12. Johanna Drucker, "Design Agency," *Dialectic* 1, no. 2 (2017), 11–16.

13. Posner, in presentation, UCLA, 2016. No online or published version exists, nor does she cite this on her resume, but the work made a vivid and important argument about supply chains and about the interface software that is an industry standard.

14. Drucker, "Design Agency."

15. See Johanna Drucker, "Performative Materiality and Theoretical Approaches to Interface," *DHQ* (*Digital Humanities Quarterly*) 7, no. 1 (Summer 2013), http://www .digitalhumanities.org/dhq/vol/7/1/000143/000143.html.

16. Lori Emerson, https://loriemerson.net/tag/digital-poetry/.

17. Drucker, "Performative Materiality."

18. The CATMA system, developed at Hamburg University by Jan Christoph Meister's team.

Appendix

1. More information on all of this and other early projects can be found in Johanna Drucker, *SpecLab* (Chicago: University of Chicago Press, 2009).

Index